Building Code Requirements for Masonry Structures
(ACI 530-95/ASCE 5-95/TMS 402-95)

Specification for Masonry Structures
(ACI 530.1-95/ASCE 6-95/TMS 602-95)

Commentary on
Building Code Requirements for Masonry Structures
(ACI 530-95/ASCE 5-95/TMS 402-95)

Commentary on
Specification for Masonry Structures
(ACI 530.1-95/ASCE 6-95/TMS 602-95)

American Concrete Institute
22400 West Seven Mile Road
Detroit, Michigan 48219

American Society of Civil Engineers
345 East 47th Street
New York, NY 10017-2398

The Masonry Society
3970 Broadway, Unit 201 D
Boulder, CO 80304

ABSTRACT

Building Code Requirement for Masonry Structures (ACI 530-95/ASCE 5-95/TMS 402-95), Specification for Masonry Structures (ACI 530.1-95/ASCE 6-95/TMS 602-95), Commentary on Building Code Requirements for Masonry Structures (ACI 530-95/ASCE 5-95/TMS 402-95) Commentary on Specification for Masonry Structures (ACI 530.1-95/ASCE 6-95/TMS 602-95) are joint efforts of the American Concrete Institute, the American Society of Civil Engineers, and The Masonry Society. The Code covers the design and construction of masonry structures while the Specification is concerned with the quality, inspection, testing and placement of materials used in construction. Some of the topics covered in the Code are: definitions, analysis and design, strength, axial loads, shear, beams, and seismic design. On the other hand, the Specification discusses subjects such as quality assurance, requirements for materials, the placing, bonding and anchoring of masonry, and the placement of grout and reinforcements. Since the Code is written as a legal document and the Specification as a master specification required by the Code, the two Commentaries present background details, committee considerations, and research data used to develop the Code and Specification.

Library of Congress Cataloging-in-Publication Data

Building code requirements for masonry structures (ACI 530-95/ASCE 5-95/TMS 402-95); Specification for masonry structures (ACI 530.1-95/ASCE 6-95/TMS 602-95); Commentary on Building code requirements for masonry structures (ACI 530-95/ASCE 5-95/TMS 402-95); Commentary on specification for masonry structures (ACI 530.1-95/ASCE 6-95/TMS 602-95).
     p.   cm.
Reported by the Masonry Standards Joint Committee.
ISBN 0-7844-0115-2
    1. Masonry—Standards—United States   I. Masonry Standards Joint Committee.
II. Title: Specification for masonry structures (ACI 530.1-95/ASCE 6-95/TMS 602-95)
III. Title: Commentary on Building code requirements for masonry structures (ACI 530-95/ASCE 595/TMS 402-95)   IV. Title: Commentary on Specification for masonry structures (ACI 530.195/ASCE 6-95/TMS 602-95)
TH1199.B85   1996   92-33528
693'.1'021873—dc20
Printed on recycled paper. 85% recovered fiber and 15% post-consumer waste.

Building Code Requirements for Masonry Structures (ACI 530-95/ASCE 5-95/TMS 402-95)

Reported by the Masonry Standards Joint Committee

James Colville
Chairman

Max L. Porter
Vice Chairman

J. Gregg Borchelt
Secretary

Maribeth S. Bradfield
Membership Secretary

Regular Members[1]:

Gene C. Abbate
Bechara E. Abboud
Bijan Ahmadi
Amde M. Amde
Richard H. Atkinson
William G. Bailey
Stuart R. Beavers
Robert J. Beiner
Frank Berg
Russell H. Brown
A. Dwayne Bryant
Kevin D. Callahan
Mario J. Catani
Robert W. Crooks
Kenneth G. Dagostino, Jr.

Gerald A. Dalrymple
Steve Dill
Russell T. Flynn
John A. Frauenhoffer
Thomas A. Gangel
Richard M. Gensert
Satyendra K. Ghosh
Clayford T. Grimm
John C. Grogan
Craig K. Haney
Gary C. Hart
Barbara Heller
Robert Hendershot
Mark B. Hogan
Thomas A. Holm

Rochelle C. Jaffe
John C. Kariotis
Richard E. Klingner
Walter Laska
L. Donald Leinweber
Hugh C. MacDonald, Jr.
Billy R. Manning
John H. Matthys
Robert McCluer
Donald G. McMican
George A. Miller
Reg Miller
Colin C. Munro
W. Thomas Munsell
Antonio Nanni

Joseph F. Neussendorfer
Joseph E. Saliba
Arturo Schultz
Matthew J. Scolforo
Daniel Shapiro
John M. Sheehan
Robert A. Speed
Ervell A. Staab
Jerry G. Stockbridge
Itzhak Tepper
Robert C. Thacker
Donald W. Vannoy
Terence A. Weigel
A. Rhett Whitlock

Associate Members[2]:

James E. Amrhein
David T. Biggs
James W. Cowie
John Chrysler
Terry M. Curtis
Walter L. Dickey
Jeffrey L. Elder
Brent A. Gabby

Hans R. Ganz
H. R. Hamilton, III
B. A. Haseltine
Edwin G. Hedstrom
A. W. Hendry
Thomas F. Herrell
Steve Lawrence
Nicholas T. Loomis

Robert F. Mast
John Melander
Raul Alamo Neihart
Robert L. Nelson
Rick Okawa
Adrian W. Page
Ruiz Lopez M. Rafael
Roscoe Reeves, Jr.

Phillip J. Samblanet
Richard C. Schumacher
John G. Tawresey
Robert D. Thomas
Dean J. Tills
Charles W. C. Yancey

SYNOPSIS

This Code covers the design and construction of masonry structures. It is written in such form that it may be adopted by reference in a general building code.

Among the subjects covered are: definitions, contract documents; quality assurance; materials; placement of embedded items; analysis and design; strength and serviceability; flexural and axial loads; shear; details and development of reinforcement; walls; columns; pilasters; beams and lintels; seismic design requirements; glass unit masonry; and veneers. An empirical design method and a prescriptive method applicable to buildings meeting specific location and construction criteria, are also included.

The quality, inspection, testing and placement of materials used in construction are covered by reference to ACI 530.1/ASCE 6/TMS 602 Specification and other standards.

Keywords: anchors (fasteners); anchorage (structural); beams; building codes; cements; clay brick; clay tile; columns; compressive strength; concrete block; concrete brick; construction; detailing; empirical design flexural strength; glass units; grout; grouting; joints; loads (forces); masonry; masonry cements; masonry load-bearing walls; masonry mortars; masonry walls; modulus of elasticity; mortars; pilasters; quality assurance; reinforced masonry; reinforcing steel; seismic requirements; shear strength; specifications; splicing; stresses; structural analysis; structural design; ties; unreinforced masonry; veneers (anchored); walls; working stress design.

[1] Regular members fully participate in Committee activities, including responding to correspondence and voting.

[2] Associate members monitor Committee activities, but do not have voting privileges.

Adopted as a standard of the American Concrete Institute November 1988, in accordance with the Institute's standardization procedure. Revised by the Institute's Expedited Standardization Procedure effective September 1, 1995. Adopted as a standard of the American Society of Civil Engineers August 1989, in accordance with the Society's standardization procedure and revised by the Society's standardization procedure effective August 1, 1995. Adopted as a standard of The Masonry Society, July 1, 1992 in accordance with the Society's standardization procedure and revised by the Society's standardization procedures effective August 1, 1995.

CONTENTS

PART 1—GENERAL

CHAPTER 1—GENERAL REQUIREMENTS

1.1—Scope

1.1.1 This Code provides minimum requirements for the structural design and construction of masonry elements consisting of masonry units bedded in mortar in any structure erected under requirements of the legally adopted general building code of which this Code forms a part. In areas without a legally adopted building code, this Code defines minimum acceptable standards of design and construction practice. Metric values shown in parentheses are provided for information only and are not part of this Code.

1.1.2 This Code supplements the general building code and shall govern in all matters pertaining to design and construction of masonry structural elements, except where this Code is in conflict with requirements in the legally adopted general building code.

1.2—Contract Documents and calculations

1.2.1 Project Drawings and project specifications for masonry structures shall identify the individual responsible for their preparation.

1.2.2 Show all Code-required drawing items on the Project Drawings, including:

(a) Name and date of issue of code and supplement to which design conforms

(b) All loads used in the design of masonry elements

(c) Specified compressive strength of masonry at stated ages or stages of construction for which masonry element is designed, except where specifically exempted by Code provisions.

(d) Specified size, grade, type and location of reinforcement, anchors and wall ties

(e) Reinforcing bars to be welded and welding requirements

(f) Size and location of all structural elements

(g) Provision for dimensional changes resulting from elastic deformation, creep, shrinkage, temperature and moisture.

1.2.3 Contract Documents shall be coordinated with design concepts and shall include an itemized program of quality assurance.

1.2.4 Calculations pertinent to design shall be filed with the drawings when required by the Building Official. When automatic data processing is used, design assumptions, program documentation and identified input and output data may be submitted in lieu of calculations.

1.3—Approval of special systems of design or construction

1.3.1 Sponsors of any system of design or construction within the scope of this Code, the adequacy of which has been shown by successful use or by analysis or test, but that does not conform to or is not covered by this Code, shall have the right to present the data on which their design is based to a board of examiners appointed by the Building Official. The board shall be composed of registered engineers and shall have authority to investigate the data so submitted, to require tests, and to formulate rules governing design and construction of such systems to meet the intent of this Code. The rules, when approved and promulgated by the Building Official, shall be of the same force and effect as the provisions of this Code.

1.4—Standards cited in this Code

1.4.1 Standards of the American Concrete Institute, the American Society of Civil Engineers, the American Society for Testing and Materials, and the American Welding Society referred to in this Code are listed below with their serial designations, including year of adoption or revision, and are declared to be part of this Code as if fully set forth in this document.

ACI 530.1-95/ ASCE 6-95/ TMS 602-95	Specification for Masonry Structures
ASCE 7-93	Minimum Design Loads for Buildings and Other Structures
ASTM C 426-70 (1993)[ε1]	Test Method for Drying Shrinkage of Concrete Block
ASTM C 476-91	Specification for Grout for Masonry
ASTM E 111-82 (1988)[ε1]	Test Method for Young's Modulus, Tangent Modulus, and Chord Modulus
ASTM E 488-90	Test Methods for Strength of Anchors in Concrete and Masonry Elements
ANSI/AWS D1.4-92	Structural Welding Code—Reinforcing Steel

CHAPTER 2—NOTATIONS AND DEFINITIONS

2.1—Notations

A_b = cross-sectional area of an anchor bolt, in.^{2} (mm^2)

A_n = net cross-sectional area of masonry, in.^{2} (mm^2)

A_p = projected area on the masonry surface of a right circular cone for anchor bolt allowable shear and tension calculations, in.^{2} (mm^2)

A_v = cross-sectional area of shear reinforcement, in.^{2} (mm^2)

A_1 = bearing area, in.^{2} (mm^2)

A_2 = effective bearing area, in.^{2} (mm^2)

A_{st} = total area of laterally tied longitudinal reinforcing steel in a reinforced masonry column or pilaster, in.^{2} (mm^2)

B_a = allowable axial force on an anchor bolt, lb (N)

B_v = allowable shear force on an anchor bolt, lb (N)

b = width of section, in. (mm)

b_a = total applied design axial force on an anchor bolt, lb (N)

b_v = total applied design shear force on an anchor bolt, lb (N)

b_w = width of wall beam, in. (mm)

D = dead load or related internal moments and forces

d = distance from extreme compression fiber to centroid of tension reinforcement, in. (mm)

d_b = nominal diameter of reinforcement, in. (mm)

d_v = actual depth of masonry in direction of shear considered, in. (mm)

E = load effects of earthquake, or related internal moments and forces

E_m = modulus of elasticity of masonry in compression, psi (MPa)

E_s = modulus of elasticity of steel, psi (MPa)

E_v = modulus of rigidity (shear modulus) of masonry, psi (MPa)

e = eccentricity of axial load, in. (mm)

F = lateral pressure of liquids, or related internal moments and forces

F_a = allowable compressive stress due to axial load only, psi (MPa)

F_b = allowable compressive stress due to flexure only, psi (MPa)

F_s = allowable tensile or compressive stress in reinforcement, psi (MPa)

F_v = allowable shear stress in masonry, psi (MPa)

f_a = calculated compressive stress in masonry due to axial load only, psi (MPa)

f_b = calculated compressive stress in masonry due to flexure only, psi (MPa)

f_g = compressive strength of grout determined in accordance with ACI 530.1/ASCE 6/TMS 602, psi (MPa)

f'_m = specified compressive strength of masonry, psi (MPa)

f_s = calculated tensile or compressive stress in reinforcement, psi (MPa)

f_v = calculated shear stress in masonry, psi (MPa)

f_y = specified yield strength of steel for reinforcement and anchors, psi (MPa)

H = lateral pressure of soil or related internal moments and forces

h = effective height of column, wall, or pilaster, in. (mm)

I = moment of inertia of masonry, in.^{4} (mm^4)

j = ratio of distance between centroid of flexural compressive forces and centroid of tensile forces to depth, d

k_c = coefficient of creep of masonry, per psi (MPa)

k_e = coefficient of irreversible moisture expansion of clay masonry

k_m = coefficient of shrinkage of concrete masonry

k_t = coefficient of thermal expansion of masonry per degree Fahrenheit (degree Celsius)

L = live load or related internal moments and forces

l = clear span between supports, in. (mm)

l_b = effective embedment length of plate, headed or bent anchor bolts, in. (mm)

l_{be} = anchor bolt edge distance measured from the surface of an anchor bolt to the nearest free edge of masonry, in. (mm)

l_d = embedment length or lap length of straight reinforcement, in.(mm)

l_e = equivalent embedment length provided by standard hooks, in. (mm)

M = maximum moment occurring simultaneously with design shear force V at the section under consideration, in.-lb (N·m)

N_v = force acting normal to shear surface, lb (N)

P = design axial load, lb (N)

P_a = allowable compressive force in reinforced masonry due to axial load, lb (N)

P_e = Euler buckling load, lb (N)

Q = first moment about the neutral axis of a section of that portion of the cross section lying between the neutral axis and extreme fiber, in.^{3} (mm^3)

r = radius of gyration, in. (mm)

s = spacing of reinforcement, in. (mm)

s_l = total linear drying shrinkage of concrete masonry units determined in accordance with ASTM C 426

T = forces and moments caused by restraint of temperature, shrinkage, and creep strains or differential movements

t = nominal thickness of wall, in. (mm)

v = shear stress, psi (MPa)

V = design shear force, lb (N)

W = wind load or related internal moments and forces

β_b = ratio of area of reinforcement cut off to total area of tension reinforcement at a section

ϕ = strength reduction factor

2.2—Definitions

Anchor—Metal rod, wire, or strap that secures masonry to its structural support.

Architect/Engineer—The architect, engineer, architectural firm, engineering firm, or architectural engineering firm issuing project drawings and specifications, or administering the work under contract specifications and drawings, or both.

Area, gross cross-sectional—The area delineated by the out-to-out dimensions of masonry in the plane under consideration.

Area, net cross-sectional—The area of masonry units, grout and mortar crossed by the plane under consideration based on out-to-out dimensions.

Backing—The wall or surface to which the veneer is secured. The backing shall be concrete, masonry, steel framing, or wood framing.

Bed joint—The horizontal layer of mortar on which a masonry unit is laid.

Building Official—The officer or other designated authority charged with the administration and enforcement of this Code, or the Building Official's duly authorized representative.

Cavity wall—A multiwythe noncomposite masonry wall with a continuous air space within the wall (with or without insulation), which is tied together with metal ties.

Collar joint—Vertical longitudinal space between wythes of masonry or between masonry wythe and back up construction, which is permitted to be filled with mortar or grout.

Column—An isolated vertical member whose horizontal dimension measured at right angles to its thickness does not exceed 3 times its thickness and whose height is at least 3 times its thickness.

Composite action—Transfer of stress between components of a member designed so that in resisting loads, the combined components act together as a single member.

Composite masonry—Multi-component masonry members acting with composite action.

Compressive strength of masonry—Maximum compressive force resisted per unit of net cross-sectional area of masonry, determined by the testing of masonry prisms or a function of individual masonry units, mortar and grout in accordance with the provisions of ACI 530.1/ASCE 6/TMS 602.

Connector—A mechanical device for securing two or more pieces, parts, or members together, including anchors, wall ties, and fasteners.

Design story drift— The difference of deflections at the top and bottom of the story under consideration, calculated by multiplying the deflections determined from an elastic analysis by the appropriate deflection amplification factor, C_d, from ASCE 7.

Diaphragm—A roof or floor system designed to transmit lateral forces to shear walls or other lateral load resisting elements.

Dimension, nominal—A nominal dimension is equal to a specified dimension plus an allowance for the joints with which the units are to be laid. Nominal dimensions are usually stated in whole numbers. Thickness is given first, followed by height and then length.

Dimensions, specified—Dimensions specified for the manufacture or construction of a unit, joint, or element.

Effective height—Clear height of a braced member between lateral supports and used for calculating the slenderness ratio of a member. Effective height for unbraced members shall be calculated.

Glass unit masonry—Nonload-bearing masonry composed of glass units bonded by mortar.

Head joint—Vertical mortar joint placed between masonry units within the wythe at the time the masonry units are laid.

Header (Bonder)—A masonry unit that connects two or more adjacent wythes of masonry.

Load, dead—Dead weight supported by a member, as defined by the general building code.

Load, live—Live load specified by the general building code.

Load, service—Load specified by the general building code.

Modulus of elasticity—Ratio of normal stress to corresponding strain for tensile or compressive stresses below proportional limit of material.

Modulus of rigidity—Ratio of unit shear stress to unit shear strain for unit shear stress below the proportional limit of the material.

Project Drawings—The drawings that, along with the Project Specifications, complete the descriptive information for constructing the Work required by the Contract Documents.

Running bond—The placement of masonry units such that head joints in successive courses are horizontally offset at least one-quarter the unit length.

Specified compressive strength of masonry, f'_m— Minimum compressive strength, expressed as force per unit of net cross-sectional area, required of the masonry used in construction by the Contract Documents, and upon which the project design is based. Whenever the quantity f'_m is under the radical sign, the square root of numerical value only is intended and the result has units of lb per in.^{2} (MPa).

Stack bond—For the purpose of this Code stack bond is other than running bond. Usually the placement of units is such that the head joints in successive courses are vertically aligned.

Stone masonry—Masonry composed of field, quarried, or cast stone units bonded by mortar.

Stone masonry, ashlar—Stone masonry composed of rectangular units having sawed, dressed, or squared bed surfaces and bonded by mortar.

Stone masonry, rubble—Stone masonry composed of irregular shaped units bonded by mortar.

Tie, lateral—Loop of reinforcing bar or wire enclosing longitudinal reinforcement.

Tie, wall—Metal connector which connects wythes of masonry walls together.

Unreinforced masonry—Masonry in which the tensile resistance of masonry is taken into consideration and the resistance of the reinforcing steel is neglected.

Veneer, adhered—Masonry veneer secured to and supported by the backing through adhesion.

Veneer, anchored—Masonry veneer secured to and supported laterally by the backing through anchors and supported vertically by the foundation or other structural elements.

Veneer, masonry—A masonry wythe which provides the exterior finish of a wall system and transfers out-of-plane load directly to a backing, but is not considered to add load resisting capacity to the wall system.

Wall—A vertical element with a horizontal length to thickness ratio greater than 3, used to enclose space.

Wall, load bearing—Wall carrying vertical loads greater than 200 lb per lineal ft (2918 N/m) in addition to its own weight.

Wall, masonry bonded hollow—A multiwythe wall built with masonry units arranged to provide an air space between the wythes and with the wythes bonded together with masonry units.

Wythe—Each continuous, vertical section of a wall, one masonry unit in thickness.

PART 2—QUALITY ASSURANCE AND CONSTRUCTION REQUIREMENTS

CHAPTER 3—GENERAL

3.1—Materials, labor and construction

3.1.1 Composition, quality, storage, handling, preparation and placement of materials, quality assurance for materials and masonry, and construction of masonry shall comply with ACI 530.1/ASCE 6/TMS 602. A quality assurance program shall be used to ensure that the constructed masonry is in conformance with the Contract Documents.

3.1.2 *Grouting, minimum spaces*—The minimum dimensions of spaces provided for the placement of grout shall be in accordance with Table 3.1.2.

Table 3.1.2—Grout space requirements

Grout type[1]	Maximum grout pour height, ft (m)	Minimum width of grout space, in. (mm)[2,3]	Minimum grout space dimensions for grouting cells of hollow units, in. x in. (mm x mm)[3,4]
Fine	1 (0.305)	$^3/_4$ (19)	1½ x 2 (38 x 51)
Fine	5 (1.52)	2 (51)	2 x 3 (51 x 76)
Fine	12 (3.66)	2½ (64)	2½ x 3 (64 x 76)
Fine	24 (7.32)	3 (76)	3 x 3 (76 x 76)
Coarse	1 (0.305)	1½ (38)	1½ x 3 (38 x 76)
Coarse	5 (1.52)	2 (51)	2½ x 3 (64 x 76)
Coarse	12 (3.66)	2½ (64)	3 x 3 (76 x 76)
Coarse	24 (7.32)	3 (76)	3 x 4 (76 x 102)

[1] Fine and coarse grouts are defined in ASTM C 476. Grout shall attain a minimum compressive strength of 2000 psi (13.8 MPa) at 28 days.

[2] For grouting between masonry wythes.

[3] Grout space dimension is the clear dimension between any masonry protusion and shall be increased by the diameters of the horizontal bars within the cross section of the grout space.

[4] Area of vertical reinforcement shall not exceed 6 percent of the area of the grout space.

3.2—Acceptance relative to strength requirements

3.2.1 *Compliance with f_m'*—Compressive strength of masonry shall be considered satisfactory if the compressive strength of each masonry wythe and grouted collar joint equals or exceeds the value of f_m'.

3.2.2 *Determination of compressive strength*

3.2.2.1 Compressive strength of masonry shall be determined in accordance with the provisions of ACI 530.1/ASCE 6/TMS 602.

CHAPTER 4—EMBEDDED ITEMS— ANCHORAGE OF MASONRY TO FRAMING AND EXISTING CONSTRUCTION

4.1—Embedded conduits, pipes, and sleeves

4.1.1 Conduits, pipes, and sleeves of any material to be embedded in masonry shall be compatible with masonry and shall comply with the following requirements.

4.1.1.1 Design shall not consider conduits, pipes, or sleeves as structurally replacing the displaced masonry.

4.1.1.2 Design shall consider the structural effects resulting from the removal of masonry to allow for the placement of pipes or conduits.

4.1.1.3 The size and location of conduits, pipes, and sleeves in masonry shall be shown on the Project Drawings. They shall be no closer than 3 diameters on center.

4.1.1.4 Maximum area of vertical conduits, pipes, or sleeves placed in masonry columns or pilasters shall not exceed 2 percent of the cross-sectional area of the column or pilaster.

4.1.1.5 Pipes shall not be embedded in masonry when:

(a) Containing liquid, gas, or vapors at temperature higher than 150° F (66° C).

(b) Under pressure in excess of 55 psi (0.38 MPa).

(c) Containing water or other liquids subject to freezing.

4.2—Anchorage of masonry to structural members, frames, and other construction

4.2.1 Anchorage of masonry to structural members, frames, and other construction shall be detailed on the Project Drawings which shall show the type, size, and location of anchor bolts. Anchor bolts shall meet the requirements of Section 5.14.

4.3—Connectors

4.3.1 Type, size, and location of connectors shall be shown or indicated on the Project Drawings.

PART 3—ANALYSIS AND DESIGN

CHAPTER 5—GENERAL ANALYSIS AND DESIGN REQUIREMENTS

5.1—Scope

5.1.1 Structures and their component members shall be designed by elastic analysis, using service loads and permissible stresses in accordance with the provisions of Chapters 5 and 8 and either Chapter 6 or 7, except as indicated under Sections 5.1.2 through 5.1.5.

5.1.2 Members that are not part of the lateral force-resisting system of the building are permitted to be designed in accordance with the provisions of Chapter 9 or Chapter 12.

5.1.3 Buildings within the limitations of Chapter 9, and their component members, are permitted to be empirically designed in accordance with the provisions of Chapter 9.

5.1.4 Glass unit masonry panels shall be empirically designed in accordance with the provisions of Chapter 11.

5.1.5 Masonry veneer and its anchors are permitted to be designed and detailed in accordance with Chapter 12 in lieu of Chapters 5, 6, 7, 8, and 9.

5.2—Loading

5.2.1 Design provisions of this Code are based on the assumption that structures shall be designed to resist all applicable loads.

5.2.2 Service loads shall be in accordance with the general building code of which this Code forms a part, with such live load reductions as are permitted in the general building code. In the absence of service loads in the general building code, the load provisions of ASCE 7 shall be used.

5.2.3 Buildings shall be provided with a structural system designed to resist wind and earthquake loads and to accommodate the effect of the resulting deformations.

5.2.4 Consideration shall be given to effects of forces and deformations due to prestressing, vibrations, impact, shrinkage, expansion, temperature changes, creep, unequal settlement of supports, and differential movement.

5.3—Load combinations

5.3.1 When the general building code does not provide load combinations, structures and members shall be designed to resist the most restrictive of the following combination of loads.

1. D
2. $D + L$
3. $D + L + (W \text{ or } E)$
4. $D + W$
5. $0.9\,D + E$
6. $D + L + (H \text{ or } F)$
7. $D + (H \text{ or } F)$
8. $D + L + T$
9. $D + T$

5.3.2 The allowable stresses in Chapters 6 and 7 are permitted to be increased by one-third when considering load combination 3, 4, or 5 of Section 5.3.1.

5.4—Design strength

5.4.1 Project Drawings shall show the specified compressive strength of masonry, f_m', for each part of the structure.

5.4.2 Each portion of the structure shall be designed based on the specified compressive strength of masonry, f_m', for that part of the work.

5.5—Material properties

5.5.1 Unless otherwise determined by test, the following moduli and coefficients shall be used in determining the effects of elasticity, temperature, moisture expansion, shrinkage, and creep.

5.5.2 *Elastic moduli*

5.5.2.1 *Steel reinforcement*

$$E_s = 29{,}000{,}000 \text{ psi } (200{,}000 \text{ MPa})$$

5.5.2.2 *Clay masonry*

(a) The design of clay masonry shall be based on the modulus of elasticity value from Table 5.5.2.2 or on the chord modulus of elasticity taken between 0.05 and 0.33 of the maximum compressive strength of each prism determined by test in accordance with the Prism Test Method, Article 1.4.B.3 of ACI 530.1/ASCE 6/TMS 602, and ASTM E 111.

(b) $E_v = 0.4\, E_m$

Table 5.5.2.2—Clay masonry

Net area compressive strength of units, psi (MPa)	Moduli of elasticity[1] E_m, psi x 10⁶ (MPa x 10³)		
	Type N mortar	Type S mortar	Type M mortar
12,000 (82.7) and greater	2.8 (19)	3.0 (21)	3.0 (21)
10,000 (68.9)	2.4 (17)	2.9 (20)	3.0 (21)
8000 (55.1)	2.0 (14)	2.4 (17)	2.8 (19)
6000 (41.3)	1.6 (11)	1.9 (13)	2.2 (15)
4000 (27.6)	1.2 (8)	1.4 (10)	1.6 (11)
2000 (13.8)	0.8 (6)	0.9 (6)	1.0 (7)

[1] Linear interpolation permitted.

5.5.2.3 *Concrete masonry*

(a) The design of concrete masonry shall be based on the modulus of elasticity value taken from Table 5.5.2.3 or on the chord modulus of elasticity taken between 0.05 and 0.33 of the maximum compressive strength of each prism determined by test in accordance with the Prism Test Method, Article 1.4.B.3 of ACI 530.1/ASCE 6/TMS 602, and ASTM E 111.

(b) $E_v = 0.4\, E_m$

Table 5.5.2.3—Concrete masonry

Net area compressive strength of units, psi (MPa)	Moduli of elasticity[1] E_m, psi x 10^6 (MPa x 10^3)	
	Type N mortar	Type M or S mortar
6000 (41.3) and greater	—	3.5 (24)
5000 (34.5)	2.8 (19)	3.2 (22)
4000 (27.6)	2.6 (18)	2.9 (20)
3000 (20.7)	2.3 (16)	2.5 (17)
2500 (17.2)	2.2 (16)	2.4 (17)
2000 (13.8)	1.8 (12)	2.2 (15)
1500 (10.3)	1.5 (10)	1.6 (11)

[1] Linear interpolation permitted.

5.5.2.4 *Grout*—Modulus of elasticity of grout shall be determined by the expression $500 f_g$.

5.5.3 Thermal expansion coefficients

5.5.3.1 *Clay masonry*

$$k_t = 4 \times 10^{-6} \text{ in./in./F } (7.2 \times 10^{-6} \text{ mm/mm/C})$$

5.5.3.2 *Concrete masonry*

$$k_t = 4.5 \times 10^{-6} \text{ in./in./F } (8.1 \times 10^{-6} \text{ mm/mm/C})$$

5.5.4 Moisture expansion coefficient of clay masonry

$$k_e = 3 \times 10^{-4} \text{ in./in. } (3 \times 10^{-4} \text{ mm/mm})$$

5.5.5 Shrinkage coefficients of concrete masonry

5.5.5.1 Masonry made of moisture controlled concrete masonry units:

$$k_m = 0.15 s_l$$

where s_t is not more than 6.5×10^{-4} in./in. (6.5×10^{-4} mm/mm)

5.5.5.2 Masonry made of non-moisture controlled concrete masonry units:

$$k_m = 0.5 s_l$$

5.5.6 Creep coefficients

5.5.6.1 *Clay masonry*

$$k_c = 0.7 \times 10^{-7}, \text{ per psi } (0.1 \times 10^{-4}, \text{ per MPa})$$

5.5.6.2 *Concrete masonry*

$$k_c = 2.5 \times 10^{-7}, \text{ per psi } (0.36 \times 10^{-4}, \text{ per MPa})$$

5.6—Deflection of beams and lintels

5.6.1 Deflection of beams and lintels due to dead plus live loads shall not exceed *l*/600 nor 0.3 in. (7.6 mm) when providing vertical support to masonry elements designed in accordance with Chapter 6 or Chapter 9.

5.7—Lateral load distribution

5.7.1 Lateral loads shall be distributed to the structural system in accordance with member stiffnesses and shall comply with the requirements of this section.

5.7.1.1 Flanges of intersecting walls designed in accordance with Section 5.13.4.2 shall be included in stiffness determination.

5.7.2 Distribution of load shall be consistent with the forces resisted by foundations.

5.7.3 Distribution of load shall include the effect of horizontal torsion of the structure due to eccentricity of wind or seismic loads resulting from the nonuniform distribution of mass.

5.8—Multiwythe walls

5.8.1 Design of walls composed of more than one wythe shall comply with the provisions of this section.

5.8.2 Composite action

5.8.2.1 Multiwythe walls designed for composite action shall have collar joints either:

(a) crossed by connecting headers, or

(b) filled with mortar or grout and connected by wall ties.

5.8.2.2 Shear stresses developed in the planes of interfaces between wythes and collar joints or within headers shall not exceed the following:

(a) mortared collar joints, 5 psi (0.034 MPa).

(b) grouted collar joints, 10 psi (0.069 MPa).

(c) headers,

$\sqrt{\text{unit compressive strength of header}}$, psi (MPa) (over net area of header).

5.8.2.3 Stresses shall be determined in accordance with the method indicated in Section 5.13.1.

5.8.2.4 Headers of wythes bonded by headers shall meet the requirements of Section 5.8.2.2 and shall be provided as follows:

(a) Headers shall be uniformly distributed and the sum of their cross-sectional areas shall be at least 4 percent of the wall surface area.

(b) Headers connecting adjacent wythes shall be embedded a minimum of 3 in. (76 mm) in each wythe.

5.8.2.5 Wythes not bonded by headers shall meet the requirements of Section 5.8.2.2 and shall be bonded by wall ties provided as follows:

Wire size	Minimum number of ties required
W1.7	one wall tie per $2^2/_3$ ft^2 (0.25 m^2) of wall
W2.8	one wall tie per $4^1/_2$ ft^2 (0.42 m^2) of wall

The maximum spacing between ties shall be 36 in. (914 mm) horizontally and 24 in. (610 mm) vertically.

Cross wires of joint reinforcement are permitted to be used as wall ties. The use of rectangular wall ties to tie walls made with any type of masonry units is permitted. The use of Z wall ties to tie walls made with other than hollow masonry units is permitted.

5.8.3 Noncomposite action
—Masonry designed for noncomposite action shall comply to the following provisions:

5.8.3.1 Each wythe shall be designed to resist individually the effects of loads imposed on it.

Unless a more detailed analysis is performed, the following requirements shall be satisfied:

(a) Collar joints shall not contain headers, grout, or mortar.

(b) Gravity loads from supported horizontal members shall be resisted by the wythe nearest to the center of span of the supported member. Any resulting bending moment about the weak axis of the wall shall be distributed to each wythe in proportion to its relative stiffness.

(c) Loads acting parallel to the plane of a wall shall be carried only by the wythe on which they are applied. Transfer of stresses from such loads between wythes shall be neglected.

(d) Loads acting transverse to the plane of a wall shall be resisted by all wythes in proportion to their relative flexural stiffness.

(e) Stresses shall be determined using the net cross-sectional area of the member or part of member under consideration, and where applicable, in accordance with the method indicated in Section 5.13.1.

(f) Specified distances between wythes shall not exceed a width of 4.5 in. (114 mm) unless a detailed wall tie analysis is performed.

5.8.3.2 Wythes of walls designed for noncomposite action shall be connected by wall ties meeting the requirements of Section 5.8.2.5, except that when cavity drips are used, the spacing indicated shall be halved or evidence shall be provided to show that their strength is equivalent to ties without drips, or by adjustable ties. In addition, ties shall be designed to transmit transverse loads from one wythe to the other.

Adjustable ties shall meet the following requirements:

(a) One tie shall be provided for each 1.77 ft^2 (0.16 m^2) of wall area.

(b) Horizontal and vertical spacing shall not exceed 16 in. (406 mm).

(c) Adjustable ties shall not be used when the misalignment of bed joints from one wythe to the other exceeds $1^1/_4$ in. (32 mm).

(d) Maximum clearance between connecting parts of the tie shall be $^1/_{16}$ in. (1.6 mm).

(e) Pintle ties shall have at least two pintle legs of wire size W2.8.

5.9—Columns

5.9.1 Design of columns shall meet the following general requirements.

5.9.1.1 Minimum side dimension shall be 8 in. (203 mm) nominal.

5.9.1.2 The ratio between the effective height and least nominal dimension shall not exceed 25.

5.9.1.3 Columns shall be designed to resist applied loading. As a minimum, columns shall be designed to resist loads with an eccentricity equal to 0.1 times each side dimension. Consider each axis independently.

5.9.1.4 Vertical column reinforcement shall not be less than 0.0025 A_n nor exceed 0.04 A_n. The minimum number of bars shall be four.

5.9.1.5 Stresses shall be determined using the method indicated in Section 5.13.1.

5.9.1.6 *Lateral ties*—Lateral ties shall conform to the following:

(a) Longitudinal reinforcement shall be enclosed by lateral ties at least $^1/_4$ in. (6.4 mm) in diameter.

(b) Vertical spacing of lateral ties shall not exceed 16 longitudinal bar diameters, 48 lateral tie bar or wire diameters, or least cross-sectional dimension of the member.

(c) Lateral ties shall be arranged such that every corner and alternate longitudinal bar shall have lateral support provided by the corner of a lateral tie with an included angle of not more than 135 deg and no bar shall be farther than 6 in. (152 mm) clear on each side along the lateral tie from such a laterally supported bar. Lateral ties shall be placed in either a mortar joint or in grout. Where longitudinal bars are located around the perimeter of a circle, a complete circular lateral tie is permitted. Lap length for circular ties shall be 48 tie diameters.

(d) Lateral ties shall be located vertically not more than one-half lateral tie spacing above the top of footing or slab in any story, and shall be spaced as provided herein to not more than one-half a lateral tie spacing below the lowest horizontal reinforcement in beam, girder, slab, or drop panel above.

(e) Where beams or brackets frame into a column from four directions, lateral ties may be terminated not more than 3 in. (76 mm) below the lowest reinforcement in the shallowest of such beams or brackets.

5.10—Pilasters

5.10.1 Walls interfacing with pilasters shall not be considered as flanges unless the provisions of Section 5.13.4.2 are met.

5.10.2 Where vertical reinforcement is provided to resist axial compressive stress, lateral ties shall meet all applicable requirements of Section 5.9.1.6.

5.10.3 Stresses shall be determined using the method indicated in Section 5.13.1.

5.11—Load transfer at horizontal connections

5.11.1 Walls, columns, and pilasters shall be designed to resist all loads, moments, and shears applied at intersections with horizontal members.

5.11.1.1 Effect of lateral deflection and translation of members providing lateral support shall be considered.

5.11.2 Devices used for transferring lateral support from members that intersect walls, columns, or pilasters shall be designed to resist the forces involved. For columns, a force of not less than 1000 lb (4448 N) shall be used.

5.12—Concentrated loads

5.12.1 For computing compressive stress f_a for walls laid in running bond, concentrated loads shall not be distributed over the length of supporting wall in excess of the length of wall equal to the width of bearing areas plus four times the thickness of the supporting wall, but not to exceed the center-to-center distance between concentrated loads.

5.12.2 Bearing stresses shall be computed by distributing the bearing load over an area determined as follows:

(a) The direct bearing area A_1 or,

(b) $A_1\sqrt{A_2/A_1}$ but not more than $2A_1$, where A_2 is the supporting surface wider than A_1 on all sides or, A_2 is the area of the lower base of the largest frustrum of a right pyramid or cone having A_1 as upper base, sloping at 45 deg from the horizontal, and wholly contained within the support. For walls in other than running bond, area A_2 shall terminate at head joints.

5.12.3 Bearing stresses shall not exceed $0.25 f'_m$.

5.13—Section properties

5.13.1 *Stress computations*

5.13.1.1 Stresses shall be computed using section properties based on the minimum net cross-sectional area of the member under consideration.

5.13.1.2 In members designed for composite action, stresses shall be computed using section properties based on the minimum transformed net cross-sectional area of the composite member. The transformed area concept for elastic analysis in which areas of dissimilar materials are transformed in accordance with relative elastic moduli ratios shall apply. Actual stresses shall be used to verify compliance with allowable stress requirements.

5.13.2 *Stiffness*—Determination of stiffness based on uncracked section is permissible. Use of the average net cross-sectional area of the member considered in stiffness computations is permitted.

5.13.3 *Radius of gyration*—Radius of gyration shall be computed using average net cross-sectional area of the member considered.

5.13.4 *Intersecting walls*

5.13.4.1 Wall intersections shall meet one of the following requirements:

(a) Design shall conform to the provisions of Section 5.13.4.2.

(b) Transfer of shear between walls shall be prevented.

5.13.4.2 Design of wall intersection.

(a) Masonry shall be in running bond.

(b) Flanges shall be considered effective in resisting applied loads.

(c) The width of flange considered effective on each side of the web shall be the lesser of 6 times the flange thickness or the actual flange on either side of the web wall.

(d) Design for shear, including the transfer of shear at interfaces, shall conform to the requirements of Section 6.5 or 7.5.

(e) The connection of intersecting walls shall conform to one of the following requirements:

  1. Fifty percent of the masonry units at the interface shall interlock.

  2. Walls shall be regularly toothed with 8 in. maximum offsets and anchored by steel connectors meeting the following requirements:

(a) Minimum size: $^1/_4$ in. x $1^1/_2$ in. x 28 in. (6.4 mm x 38 mm x 711 mm) including 2 in. (51 mm) long 90 deg bends at each end to form a U or Z shape.

(b) Maximum spacing: 4 ft (1.2 m).

  3. Intersecting bond beams shall be provided in intersecting walls at a maximum spacing of 4 ft (1.2 m) on centers. Bond beams shall be reinforced, and the area of reinforcement shall not be less than 0.1 in.^{2} per ft (211 mm^2/m) of wall. Reinforcement shall be developed on each side of the intersection.

5.14—Anchors bolts solidly grouted in masonry

5.14.1 *Test design requirements*—Except as provided in Section 5.14.2, shall be designed based on the following provisions.

5.14.1.1 Anchors shall be tested in accordance with ASTM E 488 under stresses and conditions representing intended use except that at least five tests shall be performed.

5.14.1.2 Allowable loads shall not exceed 20 percent of the average tested strength.

5.14.2 *Plate, headed, and bent bar anchor bolts*—The allowable loads for plate anchors, headed anchor bolts, and bent bar anchor bolts (J or L type) embedded in masonry shall be determined in accordance with the provisions of Sections 5.14.2.1 through 5.14.2.4.

5.14.2.1 The minimum effective embedment length shall be 4 bolt diameters, but not less than 2 in. (51 mm).

5.14.2.2 The allowable load in tension shall be the lesser of that given by Eq. (5-1) or Eq. (5-2).

$$B_a = 0.5 A_p \sqrt{f'_m} \qquad (5\text{-}1)$$

$$B_a = 0.2 A_b f_y \qquad (5\text{-}2)$$

(a) The area A_p shall be the lesser of Eq. (5-3) or Eq. (5-4). Where the projected areas of adjacent anchor bolts overlap, A_p of each bolt shall be reduced by one-half of the overlapping area. That portion of the projected area falling in an open cell or core shall be deducted from the value of A_p calculated using Eq. (5-3) or (5-4).

$$A_p = \pi l_b^2 \qquad (5\text{-}3)$$

$$A_p = \pi l_{be}^2 \qquad (5\text{-}4)$$

(b) The effective embedment length of plate or headed bolts, l_b shall be the length of embedment measured perpendicular from the surface of the masonry to the bearing surface of the plate or head of the anchor bolt.

(c) The effective embedment length of bent anchors, l_b shall be the length of embedment measured perpendicular from the surface of the masonry to the bearing surface of the bent end minus one anchor bolt diameter.

5.14.2.3 The allowable load in shear, where l_{be} equals or exceeds 12 bolt diameters, shall be the lesser of that given by Eq. (5-5) or Eq. (5-6).

$$B_v = 350 \sqrt[4]{f'_m A_b} \qquad (5\text{-}5)$$

$$B_v = 0.12 A_b f_y \qquad (5\text{-}6)$$

Where l_{be} is less than 12 bolt diameters, the value of B_v in Eq. (5-5) shall be reduced by linear interpolation to zero at an l_{be} distance of 1 in. (25 mm).

5.14.2.4 *Combined shear and tension*—Anchors in Section 5.14.2 subjected to combined shear and tension shall be designed to satisfy Eq. (5-7).

$$\frac{b_a}{B_a} + \frac{b_v}{B_v} \leq 1 \qquad (5\text{-}7)$$

5.15—Framed construction

5.15.1 Masonry walls shall not be connected to structural frames unless the connections and walls are designed to resist all interconnecting forces and to accommodate deformations.

5.16—Stack bond masonry

5.16.1 For masonry in other than running bond, the minimum area of horizontal reinforcement shall be 0.0003 times the vertical cross-sectional area of the wall. Horizontal reinforcement shall be placed in horizontal joints, or in bond beams spaced not more than 48 in. (1219 mm) on center.

CHAPTER 6—UNREINFORCED MASONRY

6.1—Scope

6.1.1 This chapter covers requirements for unreinforced masonry as defined in Section 2.2, except as otherwise indicated in Section 6.4.

6.1.2 The provisions of this chapter are to be applied in conjunction with the provisions of Chapter 5.

6.2—Stresses in reinforcement

6.2.1 The effect of stresses in reinforcement shall be neglected.

6.3—Axial compression and flexure

6.3.1 Members subjected to axial compression, flexure, or to combined axial compression and flexure shall be designed to satisfy Eq. (6-1) and Eq. (6-2).

$$\frac{f_a}{F_a} + \frac{f_b}{F_b} \leq 1 \qquad (6\text{-}1)$$

$$P \leq (^1/_4)\, P_e \qquad (6\text{-}2)$$

where:

(a) For members having an h/r ratio not greater than 99:

$$F_a = (^1/_4) f'_m \left[1 - \left(\frac{h}{140r} \right)^2 \right] \qquad (6\text{-}3)$$

(b) For members having an h/r ratio greater than 99:

$$F_a = (^1/_4) f'_m \left(\frac{70r}{h} \right)^2 \qquad (6\text{-}4)$$

(c) $$F_b = (^1/_3) f'_m \qquad (6\text{-}5)$$

(d) $$P_e = \frac{\pi^2 E_m I}{h^2} \left(1 - 0.577 \frac{e}{r} \right)^3 \qquad (6\text{-}6)$$

6.3.1.1 Allowable tensile stresses due to flexure transverse to the plane of masonry member shall be in accordance with the values in Table 6.3.1.1.

Table 6.3.1.1—Allowable flexural tension for clay and concrete masonry, psi (MPa)

	Mortar types			
	Portland cement/lime		Masonry cement and air entrained portland cement/lime	
Masonry type	M or S	N	M or S	N
Normal to bed joints				
Solid units	40 (0.28)	30 (0.21)	24 (0.17)	15 (0.10)
Hollow units[1]				
Ungrouted	25 (0.17)	19 (0.13)	15 (0.10)	9 (0.06)
Fully grouted	68 (0.47)	58 (0.40)	41 (0.28)	26 (0.18)
Parallel to bed joints in running bond				
Solid units	80 (0.55)	60 (0.41)	48 (0.33)	30 (0.21)
Hollow units				
Ungrouted and partially grouted	50 (0.35)	38 (0.26)	30 (0.21)	19 (0.13)
Fully grouted	80 (0.55)	60 (0.41)	48 (0.33)	30 (0.21)

[1] For partially grouted masonry, allowable stresses shall be determined on the basis of linear interpolation between hollow units that are fully grouted or ungrouted and hollow units based on amount of grouting.

6.4—Axial tension

6.4.1 The tensile strength of masonry shall be neglected in design when the masonry is subjected to axial tension forces.

6.5—Shear

6.5.1 Shear stresses due to forces acting in the direction considered shall be computed in accordance with Section 5.13.1 and determined by Eq. (6-7).

$$f_v = \frac{VQ}{Ib} \qquad (6\text{-}7)$$

6.5.2 In-plane shear stresses shall not exceed any of:

(a) $1.5\sqrt{f'_m}$

(b) 120 psi (0.83 MPa)

(c) $v + 0.45\, N_v/A_n$

where v:

= 37 psi (0.26 MPa) for masonry in running bond that is not grouted solid, or

= 37 psi (0.26 MPa) for masonry in other than running bond with open end units that are grouted solid, or

= 60 psi (0.41 MPa) for masonry in running bond that is grouted solid.

(d) 15 psi (0.10 MPa) for masonry in other than running bond with other than open end units that are grouted solid.

6.5.3 Shear stresses shall not exceed the requirements of Section 5.8.2.2 at interfaces between wythes and filled collar joints or between wythes and headers.

CHAPTER 7—REINFORCED MASONRY

7.1—Scope

7.1.1 This chapter covers requirements for the design of structures neglecting the contribution of tensile strength of masonry except as provided in Section 7.5.

7.1.2 The provisions of this chapter are to be applied in conjunction with the general requirements of Chapter 5.

7.2—Steel reinforcement

7.2.1 *Allowable stresses*

7.2.1.1 *Tension*—Tensile stress in reinforcement shall not exceed the following:

    (a)   Grade 40 or Grade 50 reinforcement
    .... 20,000 psi (138 MPa)

    (b)   Grade 60 reinforcement
    .... 24,000 psi (165 MPa)

    (c)   Wire joint reinforcement
    .... 30,000 psi (207 MPa)

7.2.1.2 *Compression*

    (a)   The compressive resistance of steel reinforcement shall be neglected unless lateral reinforcement is provided in compliance with the requirements of Section 5.9.1.6.

    (b)   Compressive stress in reinforcement shall not exceed the lesser of $0.4 f_y$ or 24,000 psi (165 MPa).

7.3—Axial compression and flexure

7.3.1 Members subjected to axial compression, flexure or to combined axial compression and flexure shall be designed in compliance with the following requirements:

7.3.2 *Allowable forces and stresses*

7.3.2.1 The compressive force in reinforced masonry due to axial load only shall not exceed that given by Eq. (7-1) or Eq. (7-2).

    (a)   For members having and h/r ratio not greater than 99:

$$P_a = (0.25 f'_m A_n + 0.65 A_{st} F_s)\left[1 - \left(\frac{h}{140\,r}\right)^2\right] \quad (7\text{-}1)$$

    (b)   For members having an h/r ratio greater than 99:

$$P_a = (0.25 f'_m A_n + 0.65 A_{st} F_s)\left(\frac{70\,r}{h}\right)^2 \quad (7\text{-}2)$$

7.3.2.2 The compressive stress in masonry due to flexure or due to flexure in combination with axial load shall not exceed $(^1/_3)f'_m$ provided the calculated compressive stress due to the axial load component, f_a, does not exceed the allowable stress, F_a, in Section 6.3.1.

7.3.3 *Effective compressive width per bar*

7.3.3.1 In running bond masonry and masonry in other than running bond, with bond beams spaced not more than 48 in. center-to-center, the width of the compression area used in stress calculations shall not exceed the least of:

    (a)   Center-to-center bar spacing.
    (b)   Six times the wall thickness.
    (c)   72 in. (1829 mm).

7.3.3.2 In masonry in other than running bond with bond beams spaced more than 48 in. (1219 mm) center-to-center, the width of the compression area used in stress calculations shall not exceed the length of the masonry unit.

7.3.4 *Beams*

7.3.4.1 Span length of members not built integrally with supports shall be taken as the clear span plus depth of member, but need not exceed the distance between centers of supports.

7.3.4.2 In analysis of members that are continuous over supports for determination of moments, span length shall be taken as the distance between centers of supports.

7.3.4.3 Length of bearing of beams on their supports shall be a minimum of 4 in. (102 mm) in the direction of span.

7.3.4.4 The compression face of beams shall be laterally supported at a maximum spacing of 32 times the beam thickness.

7.3.4.5 Deflection of beams shall meet the requirements of Section 5.6.

7.4—Axial tension

7.4.1 Axial tension shall be resisted entirely by steel reinforcement.

7.5—Shear

7.5.1 Members which are not subjected to flexural tension shall be designed in accordance with the requirements of Section 6.5 or shall be designed in accordance with the following:

7.5.1.1 Reinforcement shall be provided in accordance with the requirements of Section 7.5.3.

7.5.1.2 The calculated shear stress, f_v, shall not exceed F_v, where F_v is determined in accordance with Section 7.5.2.3.

7.5.2 Members subjected to flexural tension shall be reinforced to resist the tension and shall be designed in accordance with the following:

7.5.2.1 Calculated shear stress in the masonry shall be determined by the relation:

$$f_v = \frac{V}{bd} \quad (7\text{-}3)$$

7.5.2.2 Where reinforcement is not provided to resist all of the calculated shear, f_v shall not exceed F_v, where:

    (a)   for flexural members

$$F_v = \sqrt{f'_m} \quad (7\text{-}4)$$

but shall not exceed 50 psi (0.35 MPa).

    (b)   for shear walls
    where
    $M/Vd < 1$,

$$F_v = (1/3)[4 - (M/Vd)]\sqrt{f'_m} \quad (7\text{-}5)$$

but shall not exceed $80 - 45(M/Vd)$ psi

where
$M/Vd \geq 1$,

$$F_v = \sqrt{f'_m} \tag{7-6}$$

but shall not exceed 35 psi (0.24 MPa).

7.5.2.3 Where shear reinforcement is provided in accordance with Section 7.5.3 to resist all of the calculated shear, f_v shall not exceed F_v, where:

(a)   for flexural members

$$F_v = 3.0 \sqrt{f'_m} \tag{7-7}$$

but shall not exceed 150 psi (1.03 MPa).

(b)   for shear walls

where
$M/Vd < 1$,

$$F_v = (^1/_2)[4 - (M/Vd)]\sqrt{f'_m} \tag{7-8}$$

but shall not exceed $120 - 45(M/Vd)$ psi

where
$M/Vd \geq 1$,

$$F_v = 1.5 \sqrt{f'_m} \tag{7-9}$$

but shall not exceed 75 psi (0.52 MPa).

7.5.2.4 The ratio M/Vd shall always be taken as a positive number.

7.5.3   Minimum area of shear reinforcement required by Section 7.5.1 or Section 7.5.2.3 shall be determined by the following:

$$A_v = \frac{Vs}{F_s d} \tag{7-10}$$

7.5.3.1 Shear reinforcement shall be provided parallel to the direction of applied shear force.   Spacing of shear reinforcement shall not exceed the lesser of $d/2$ or 48 in. (1219 mm).

7.5.3.2 Reinforcement shall be provided perpendicular to the shear reinforcement and shall be at least equal to one-third A_v.   The reinforcement shall be uniformly distributed and shall not exceed a spacing of 8 ft (2.4 m).

7.5.4   In composite masonry walls, shear stresses developed in the planes of interfaces between wythes and filled collar joints or between wythes and headers shall meet the requirements of Section 5.8.2.2.

7.5.5   In cantilever beams, the maximum shear shall be used.   In noncantilever beams, the maximum shear shall be used except that sections located within a distance $d/2$ from the face of support shall be designed for the same shear as that computed at a distance $d/2$ from the face of support when the following conditions are met:

(a)   support reaction, in direction of applied shear force, introduces compression into the end regions of member, and

(b)   no concentrated load occurs between face of support and a distance $d/2$ from face.

CHAPTER 8—DETAILS OF REINFORCEMENT

8.1—Scope

8.1.1 This chapter covers requirements for design of details of reinforcement.

8.1.2 Details of reinforcement shall be shown or covered by notes on the Project Drawings.

8.1.3 Reinforcing bars shall be embedded in grout.

8.2—Size of reinforcement

8.2.1 The maximum size of reinforcement used in masonry shall be No. 11.

8.2.2 The diameter of reinforcement shall not exceed one-half the least clear dimension of the cell, bond beam, or collar joint in which it is placed. (See Section 3.1.2.)

8.2.3 Longitudinal and cross wires of joint reinforcement shall have a minimum wire size of W1.1 and a maximum wire size of one-half the joint thickness.

8.3—Placement limits for reinforcement

8.3.1 The clear distance between parallel bars shall not be less than the nominal diameter of the bars, nor less than 1 in. (25 mm).

8.3.2 In columns and pilasters, the clear distance between vertical bars shall not be less than one and one-half times the nominal bar diameter, nor less than $1^{1}/_{2}$ in. (38 mm).

8.3.3 The clear distance limitations between bars required in Sections 8.3.1 and 8.3.2 shall also apply to the clear distance between a contact lap splice and adjacent splices or bars.

8.3.4 Groups of parallel reinforcing bars bundled in contact to act as a unit shall be limited to two in any one bundle. Individual bars in a bundle cut off within the span of a member shall terminate at points at least 40 bar diameters apart.

8.3.5 Reinforcement embedded in grout shall have a thickness of grout between the reinforcement and masonry units not less than $^{1}/_{4}$ in. (6.4 mm) for fine grout or $^{1}/_{2}$ in. (13 mm) for coarse grout.

8.4—Protection for reinforcement

8.4.1 Reinforcing bars shall have a masonry cover not less than the following:

(a) Masonry face exposed to earth or weather: 2 in. (51 mm) for bars larger than No. 5; $1^{1}/_{2}$ in. (38 mm) for No. 5 bars or smaller.

(b) Masonry not exposed to earth or weather: $1^{1}/_{2}$ in. (38 mm).

8.4.2 Longitudinal wires of joint reinforcement shall be fully embedded in mortar or grout with a minimum cover of $^{5}/_{8}$ in. (16 mm) when exposed to earth or weather and $^{1}/_{2}$ in. (13 mm) when not exposed to earth or weather. Joint reinforcement in masonry exposed to earth or weather shall be corrosion resistant or protected from corrosion by coating. (See Section 3.1.1.)

8.4.3 Wall ties, anchors and inserts shall be protected from corrosion, except anchor bolts not exposed to the weather or moisture.

8.5—Development of reinforcement embedded in grout

8.5.1 *General*—The calculated tension or compression in the reinforcement at each section shall be developed on each side of the section by embedment length, hook or mechanical device or a combination thereof. Hooks shall not be used to develop bars in compression.

8.5.2 *Embedment of bars and wires in tension*—The embedment length of bars and wire shall be determined by Eq. (8-1), but shall not be less than 12 in. (305 mm) for bars and 6 in. (152 mm) for wire.

$$l_d = 0.0015d_b F_s \qquad (8\text{-}1)$$

When epoxy-coated bars are used, development length determined by Eq. (8-1) shall be increased by 50 percent.

8.5.3 *Embedment of flexural reinforcement*

8.5.3.1 *General*

(a) Tension reinforcement is permitted to be developed by bending across the neutral axis of the member to be anchored or made continuous with reinforcement on the opposite face of the member.

(b) Critical sections for development of reinforcement in flexural members are at points of maximum steel stress and at points within the span where adjacent reinforcement terminates, or is bent.

(c) Reinforcement shall extend beyond the point at which it is no longer required to resist flexure for a distance equal to the effective depth of the member or $12d_b$, whichever is greater, except at supports of simple spans and at the free end of cantilevers.

(d) Continuing reinforcement shall extend a distance l_d beyond the point where bent or terminated tension reinforcement is no longer required to resist flexure as required by Section 8.5.2.

(e) Flexural reinforcement shall not be terminated in a tension zone unless one of the following conditions is satisfied:

1. Shear at the cutoff point does not exceed two-thirds of the allowable shear at the section considered.

2. Stirrup area in excess of that required for shear is provided along each terminated bar or wire over a distance from the termination point equal to three-fourths the effective depth of the member. Excess stirrup area, A_v, shall not be less than $60\, b_w s/f_y$. Spacing s shall not exceed $d/(8\, \beta_b)$.

3. Continuous reinforcement provides double the area required for flexure at the cutoff point and shear does not exceed three-fourths the allowable shear at the section considered.

(f) Adequate anchorage complying with Section 8.5.2 shall be provided for tension reinforcement in flexural members where reinforcement stress is not directly proportional to moment, such as corbels, deep flexural members, or members in which tension reinforcement is not parallel to compression face.

8.5.3.2 *Development of positive moment reinforcement*—When a wall or other flexural member is part of a primary lateral load resisting system, at least 25 percent of the positive moment reinforcement shall extend into the support and be anchored to develop a stress equal to the F_s in tension.

8.5.3.3 *Development of negative moment reinforcement*

(a) Negative moment reinforcement in a continuous, restrained, or cantilever member shall be anchored in or through the supporting member in accordance with the provisions of Section 8.5.1.

(b) At least one-third of the total reinforcement provided for moment at a support shall extend beyond the point of inflection the greater distance of the effective depth of the member or one-sixteenth of the span.

8.5.4 *Standard hooks*

8.5.4.1 The term standard hook as used in this Code shall mean one of the following:

(a) A 180 deg turn plus extension of at least 4 bar diameters but not less than $2^1/_2$ in. (64 mm) at free end of bar.

(b) A 90 deg turn plus extension of at least 12 bar diameters at free end of bar.

(c) For stirrup and tie anchorage only, either a 90 deg or a 135 deg turn plus an extension of at least 6 bar diameters at the free end of the bar.

8.5.5 *Minimum bend diameter for reinforcing bars*

8.5.5.1 The diameter of bend measured on the inside of reinforcing bars, other than for stirrups and ties, shall not be less than values specified in Table 8.5.5.1.

Table 8.5.5.1—Minimum diameters of bend

Bar size	Minimum diameter
No. 3 through No. 7	(Grade 40) 5 bar diameters
No. 3 through No. 8	(Grade 50 or 60) 6 bar diameters
No. 9, No. 10, and No. 11	(Grade 50 or 60) 8 bar diameters

8.5.5.2 Standard hooks in tension shall be considered to develop an equivalent embedment length, l_e, equal to 11.25 d_b.

8.5.5.3 The effect of hooks for bars in compression shall be neglected in design computations.

8.5.6 *Development of shear reinforcement*

8.5.6.1 *Bar and wire reinforcement*

(a) Shear reinforcement shall extend to a distance d from the extreme compression face and shall be carried as close to the compression and tension surfaces of the member as cover requirements and the proximity of other reinforcement permit. Shear reinforcement shall be anchored at both ends for its calculated stress.

(b) The ends of single leg or U-stirrups shall be anchored by one of the following means:

1. A standard hook plus an effective embedment of $0.5\ l_d$. The effective embedment of a stirrup leg shall be taken as the distance between the middepth of the member $d/2$ and the start of the hook (point of tangency).

2. For No. 5 bar and D31 wire, and smaller,

bending around longitudinal reinforcement through at least 135 deg plus an embedment of $0.33\ l_d$. The $0.33\ l_d$ embedment of a stirrup leg shall be taken as the distance between middepth of member $d/2$ and start of hook (point of tangency).

(c) Between the anchored end, each bend in the continuous portion of a transverse U-stirrup shall enclose a longitudinal bar.

(d) Longitudinal bars bent to act as shear reinforcement, where extended into a region of tension, shall be continuous with longitudinal reinforcement and, where extended into a region of compression, shall be developed beyond middepth of the member $d/2$.

(e) Pairs of U-stirrups or ties placed to form a closed unit shall be considered properly spliced when length of laps are $1.7\ l_d$. In grout at least 18 in. (457 mm) deep, such splices with $A_v f_y$ not more than 9000 lb (40,000 N) per leg may be considered adequate if legs extend the full available depth of grout.

8.5.6.2 *Welded wire fabric*

(a) For each leg of welded wire fabric forming simple U-stirrups, there shall be either:

1. Two longitudinal wires spaced at a 2 in. (51 mm) spacing along the member at the top of the U, or

2. One longitudinal wire located not more than $d/4$ from the compression face and a second wire closer to the compression face and spaced not less than 2 in. (51 mm) from the first wire. The second wire shall be located on the stirrup leg beyond a bend, or on a bend with an inside diameter of bend not less than $8d_b$.

(b) For each end of a single leg stirrup of welded smooth or deformed wire fabric, there shall be two longitudinal wires spaced a minimum of 2 in. (51 mm) with the inner wire placed at a distance at least $d/4$ or 2 in. (51 mm) from middepth of member $d/2$. Outer longitudinal wire at tension face shall not be farther from the face than the portion of primary flexural reinforcement closest to the face.

8.5.7 *Splices of reinforcement*—Lap splices, welded splices, or mechanical connections are permitted in accordance with the provisions of this section. All welding shall conform to AWS D1.4.

8.5.7.1 *Lap splices*

(a) The minimum length of lap for bars in tension or compression shall be determined by Eq. (8-2), but not less than 12 in. (305 mm).

$$l_d = 0.002 d_b F_s \qquad (8-2)$$

When epoxy-coated bars are used, lap length determined by Eq. (8-2) shall be increased by 50 percent.

(b) Bars spliced by noncontact lap splices shall not be spaced transversely farther apart than one-fifth the required length of lap nor more than 8 in. (203 mm).

8.5.7.2 *Welded splices*—Welded splices shall have the bars butted and welded to develop in tension at least 125 percent of the specified yield strength of the bar.

8.5.7.3 *Mechanical connections*—Mechanical connections shall have the bars connected to develop in tension or compression, as required, at least 125 percent of the specified yield strength of the bar.

8.5.7.4 *End bearing splices*

(a) In bars required for compression only, the transmission of compressive stress by bearing of square cut ends held in concentric contact by a suitable device is permitted.

(b) Bar ends shall terminate in flat surfaces within $1\frac{1}{2}$ deg of a right angle to the axis of the bars and shall be fitted within 3 deg of full bearing after assembly.

(c) End bearing splices shall be used only in members containing closed ties, closed stirrups, or spirals.

CHAPTER 9—EMPIRICAL DESIGN OF MASONRY

9.1—Scope

9.1.1 This chapter covers masonry buildings and masonry elements which are designed in accordance with empirical requirements in lieu of the requirements of Chapters 5, 6, 7, and 8, except as specifically stated herein.

9.1.2 *Limitations*

9.1.2.1 *Seismic*—Empirical requirements shall not apply to the design or construction of masonry for buildings, parts of buildings or other structures in Seismic Performance Category D or E as defined in ASCE 7, and shall not apply to the design of the lateral force-resisting system for structures in Seismic Performance Category B or C.

9.1.2.2 *Wind*—Empirical requirements shall not apply to the design or construction of masonry for buildings, parts of buildings, or other structures to be located in areas where the velocity pressure exceeds 25 lb/ft^2 (1190 Pa) as given in ASCE 7.

9.1.2.3 *Other horizontal loads*—Empirical requirements shall not apply to structures resisting horizontal loads other than permitted wind or seismic loads or foundation walls as provided in Section 9.6.3.

9.1.2.4 *Glass unit masonry*—The provisions of Chapter 9 do not apply to glass unit masonry.

9.2—Height

9.2.1 Buildings relying on masonry walls as part of their lateral load resisting system shall not exceed 35 ft (10.7 m) in height.

9.3—Lateral stability

9.3.1 *Shear walls*—Where the structure depends upon masonry walls for lateral stability, shear walls shall be provided parallel to the direction of the lateral forces resisted.

9.3.1.1 Minimum nominal thickness of masonry shear walls shall be 8 in. (203 mm).

9.3.1.2 In each direction in which shear walls are required for lateral stability, the minimum cumulative length of shear walls provided shall be 0.4 times the long dimension of the building. Cumulative length of shear walls shall not include openings.

9.3.1.3 Maximum spacing of masonry shear walls shall not exceed the ratio listed in Table 9.3.1.

Table 9.3.1—Shear wall spacing requirements

Floor or roof construction	Maximum ratio of shear wall spacing: shear wall length
Cast-in-place concrete	5:1
Precast concrete	4:1
Metal deck with concrete fill	3:1
Metal deck with no fill	2:1
Wood diaphragm	2:1

9.3.2 *Roofs*—The roof construction shall be designed so as not to impart out-of-plane lateral thrust to the walls under roof gravity load.

9.4—Compressive stress requirements

9.4.1 Compressive stresses in masonry due to vertical dead plus live loads (excluding wind or seismic loads) shall be determined in accordance with Section 9.4.2.1. Dead and live loads shall be in accordance with the general building code of which this Code forms a part, with such live load reductions as are permitted in the general building code.

9.4.2 The compressive stresses in masonry shall not exceed the values given in Table 9.4.2. Stress shall be calculated based on actual rather than nominal dimensions.

9.4.2.1 Calculated compressive stresses for single wythe walls and for multiwythe composite masonry walls shall be determined by dividing the design load by the gross cross-sectional area of the member. The area of openings, chases, or recesses in walls shall not be included in the gross cross-sectional area of the wall.

9.4.2.2 *Multiwythe walls*—The allowable stress shall be as given in Table 9.4.2 for the weakest combination of the units and mortar used in each wythe.

9.5—Lateral support

9.5.1 Masonry walls shall be laterally supported in either the horizontal or the vertical direction at intervals not exceeding those given in Table 9.5.1.

Table 9.5.1—Wall lateral support requirements

Construction	Maximum l/t or h/t
Bearing walls	
Solid units or fully grouted	20
All other	18
Nonbearing walls	
Exterior	18
Interior	36

In computing the ratio for multiwythe walls, use the following thickness:
1. The nominal wall thicknesses for solid walls and for hollow walls bonded with masonry headers (Section 9.7.2).
2. The sum of the nominal thicknesses of the wythes for non-composite walls connected with wall ties (Section 9.7.3).

9.5.1.1 Except for parapets, the ratio of height to nominal thickness for cantilever walls shall not exceed 6 for solid masonry or 4 for hollow masonry. For parapets see Section 9.6.4.

9.5.2 Lateral support shall be provided by cross walls, pilasters, buttresses, or structural frame members when the limiting distance is taken horizontally, or by floors, roofs acting as diaphragms or structural frame members when the limiting distance is taken vertically.

Table 9.4.2 —Allowable compressive stresses for empirical design of masonry

Construction; compressive strength of unit, gross area, psi (MPa)	Allowable compressive stresses[1] gross cross-sectional area, psi (MPa)	
	Type M or S mortar	Type N mortar
Solid masonry of brick and other solid units of clay or shale; sand-lime or concrete brick:		
8000 (55.1) or greater	350 (2.4)	300 (2.1)
4500 (31.0)	225 (1.6)	200 (1.4)
2500 (17.2)	160 (1.1)	140 (0.97)
1500 (10.3)	115 (0.79)	100 (0.69)
Grouted masonry, of clay or shale; sand-lime or concrete:		
4500 (31.0) or greater	225 (1.6)	200 (1.4)
2500 (17.2)	160 (1.1)	140 (0.97)
1500 (8.3)	115 (0.79)	100 (0.69)
Solid masonry of solid concrete masonry units:		
3000 (20.7) or greater	225 (1.6)	200 (1.4)
2000 (13.8)	160 (1.1)	140 (0.97)
1200 (8.3)	115 (0.79)	100 (0.69)
Masonry of hollow load bearing units:		
2000 (13.8) or greater	140 (0.97)	120 (0.83)
1500 (10.3)	115 (0.79)	100 (0.69)
1000 (6.9)	75 (0.52)	70 (0.48)
700 (4.8)	60 (0.41)	55 (0.38)
Hollow walls (noncomposite masonry bonded[2])		
Solid units:		
2500 (17.2) or greater	160 (1.1)	140 (0.97)
1500 (10.3)	115 (0.79)	100 (0.69)
Hollow units	75 (0.52)	70 (0.48)
Stone ashlar masonry:		
Granite	720 (5.0)	640 (4.4)
Limestone or marble	450 (3.1)	400 (2.8)
Sandstone or cast stone	360 (2.5)	320 (2.2)
Rubble stone masonry		
Coursed, rough, or random	120 (0.83)	100 (0.69)

[1] Linear interpolation for determining allowable stressses for masonry units having compressive strengths which are intermediate between those given in the table is permitted.

[2] Where floor and roof loads are carried upon one wythe, the gross cross-sectional area is that of the wythe under load; if both wythes are loaded, the gross cross-sectional area is that of the wall minus the area of the cavity between the wythes. Walls bonded with metal ties shall be considered as noncomposite walls unless collar joints are filled with mortar or grout.

9.6—Thickness of masonry

9.6.1 The nominal thickness of masonry walls shall conform to the following requirements.

9.6.2 *Thickness of walls*

9.6.2.1 *Minimum thickness*—The minimum thickness of masonry bearing walls more than one story high shall be 8 in (203 mm). Bearing walls of one story buildings shall not be less than 6 in. (152 mm) thick.

9.6.2.2 *Rubble stone walls*—The minimum thickness of rough or random or coursed rubble stone walls shall be 16 in.

9.6.2.3 *Change in thickness*—Where walls of masonry of hollow units or masonry bonded hollow walls are decreased in thickness a course or courses of solid masonry shall be interposed between the wall below and the thinner wall above, or special units or construction shall be used to transmit the loads from face shells or wythes above to those below.

9.6.3 *Foundation walls*

9.6.3.1 Foundation walls shall comply with the requirements of Table 9.6.3.1. The provisions of Table 9.6.3.1 are applicable when: (1) the foundation wall does not exceed 8 ft. (2.44m) in height between lateral supports; (2) the terrain surrounding foundation walls is graded so as to drain surface water away from foundation walls; (3) backfill is drained to remove ground water away from foundation walls; (4) lateral support is provided at the top of foundation walls prior to backfilling; (5) the length of foundation walls between perpendicular masonry walls or pilasters is a maximum of 3 times the basement wall height; (6) the backfill is granular and soil conditions in the area are non-expansive; and (7) masonry is laid in running bond using Type M or S mortar.

Table 9.6.3.1—Foundation wall construction

Wall construction	Nominal wall thickness, in. (mm)	Maximum depth of unbalanced backfill, ft (m)
Hollow unit masonry	8 (203)	5 (1.53)
	10 (254)	6 (1.83)
	12 (305)	7 (2.14)
Solid unit masonry	8 (203)	5 (1.53)
	10 (254)	7 (2.14)
	12 (305)	7 (2.14)
Fully grouted masonry	8 (203)	7 (2.14)
	10 (254)	8 (2.44)
	12 (305)	8 (2.44)

9.6.3.2 Where the requirements of 9.6.3.1 are not met, foundation walls shall be designed in accordance with Chapters 5 and 6 or Chapters 5, 7 and 8.

9.6.4 *Parapet walls*—Parapet walls shall be at least 8 in. (203 mm) thick and their height shall not exceed 3 times their thickness.

9.7—Bond

9.7.1 *General*—The facing and backing of multiple wythe masonry walls shall be bonded in accordance with Section 9.7.2, 9.7.3, or 9.7.4.

9.7.2 *Bonding with masonry headers*

9.7.2.1 *Solid units*—Where the facing and backing (adjacent wythes) of solid masonry construction are bonded by means of masonry headers, no less than 4 percent of the wall surface of each face shall be composed of headers extending not less than 3 in. (76 mm) into the backing. The distance between adjacent full-length headers shall not exceed 24 in. (610 mm) either vertically or horizontally. In walls in which a single header does not extend through the wall, headers from the opposite sides shall overlap at least 3 in. (76 mm), or headers from opposite sides shall be covered with another header course overlapping the header below at least 3 in. (76 mm).

9.7.2.2 *Hollow units*—Where two or more hollow units are used to make up the thickness of a wall, the stretcher courses shall be bonded at vertical intervals not exceeding 34 in. (864 mm) by lapping at least 3 in. (76 mm) over the unit below, or by lapping at vertical intervals not exceeding 17 in. (432 mm) with units which are at least 50 percent greater in thickness than the units below.

9.7.3 *Bonding with wall ties*

9.7.3.1 When the facing and backing (adjacent wythes) of masonry walls are bonded with wire size W2.8 wall ties or metal wire of equivalent stiffness embedded in the horizontal mortar joints, there shall be at least one metal tie for each $4\frac{1}{2}$ ft^2 (0.42 m^2) of wall area. Ties in alternate courses shall be staggered, the maximum vertical distance between ties shall not exceed 24 in. (610 mm), and the maximum horizontal distance shall not exceed 36 in. (914 mm). Rods or ties bent to rectangular shape shall be used with hollow masonry units laid with the cells vertical. In other walls the ends of ties shall be bent to 90 deg angles to provide hooks no less than 2 in. (51 mm) long. Additional bonding ties shall be provided at all openings, spaced not more than 3 ft (0.92 m) apart around the perimeter and within 12 in. (305 mm) of the opening.

9.7.3.2 *Bonding with prefabricated joint reinforcement*—Where the facing and backing (adjacent wythes) of masonry are bonded with prefabricated joint reinforcement, there shall be at least one cross wire serving as a tie for each $2\frac{2}{3}$ ft^2 (0.25 m^2) of wall area. The vertical spacing of the joint reinforcement shall not exceed 24 in. (610 mm). Cross wires on prefabricated joint reinforcement shall be not smaller than wire size W1.7. The longitudinal wires shall be embedded in the mortar.

9.7.4 *Natural or cast stone*

9.7.4.1 *Ashlar masonry*—In ashlar masonry, bonder units, uniformly distributed, shall be provided to the extent of not less than 10 percent of the wall area. Such bonder units shall extend not less than 4 in. (102 mm) into the backing wall.

9.7.4.2 *Rubble stone masonry*—Rubble stone masonry 24 in. (610 mm) or less in thickness shall have bonder units with a maximum spacing of 3 ft (0.92 m) vertically and 3 ft (0.92 m) horizontally, and if the masonry is of greater thickness than 24 in. (610 mm), shall have one bonder unit for each 6 ft^2 (0.56 m^2) of wall surface on both sides.

9.7.5 *Longitudinal bond*

9.7.5.1 Each wythe of masonry shall be laid in running bond, or the masonry walls shall be reinforced longitudinally as required in Section 9.7.5.2.

9.7.5.2 Where masonry is laid in other than running bond, reinforcement shall be provided in accordance with the provisions of Section 5.16.

9.8—Anchorage

9.8.1 *General*—Masonry elements shall be anchored in accordance with this section.

9.8.2 *Intersecting walls*—Masonry walls depending upon one another for lateral support shall be anchored or bonded at locations where they meet or intersect by one of the

following methods:

9.8.2.1 Fifty percent of the units at the intersection shall be laid in an overlapping masonry bonding pattern, with alternate units having a bearing of not less than 3 in. (76 mm) on the unit below.

9.8.2.2 Walls shall be anchored by steel connectors having a minimum section of $\frac{1}{4}$ in. (6.4 mm) by $1\frac{1}{2}$ in. (38 mm) with ends bent up at least 2 in. (51 mm), or with cross pins to form anchorage. Such anchors shall be at least 24 in. (610 mm) long and the maximum spacing shall be 4 ft (1.22 m).

9.8.2.3 Walls shall be anchored by joint reinforcement spaced at a maximum distance of 8 in. (203 mm). Longitudinal wires of such reinforcement shall be at least wire size W1.7 and shall extend at least 30 in. (762 mm) in each direction at the intersection.

9.8.2.4 Interior nonload-bearing walls shall be anchored at their intersection, at vertical intervals of not more than 16 in. (406 mm) with joint reinforcement or $\frac{1}{4}$ in. (6.4 mm) mesh galvanized hardware cloth.

9.8.2.5 Other metal ties, joint reinforcement or anchors, if used, shall be spaced to provide equivalent area of anchorage to that required by this section.

9.8.3 *Floor and roof anchorage*—Floor and roof diaphragms providing lateral support to masonry shall be connected to the masonry by one of the following methods:

9.8.3.1 Wood floor joists bearing on masonry walls shall be anchored to the wall at intervals not to exceed 6 ft (1.8 m) by metal strap anchors. Joists parallel to the wall shall be anchored with metal straps spaced not more than 6 ft (1.8 m) on centers extending over or under and secured to at least 3 joists. Blocking shall be provided between joists at each strap anchor.

9.8.3.2 Steel floor joists bearing on masonry walls shall be anchored to the wall with $\frac{3}{8}$ in. (9.5 mm) round bars, or their equivalent, spaced not more than 6 ft (1.83 m) on center. Where joists are parallel to the wall, anchors shall be located at joist bridging.

9.8.3.3 Roof diaphragms shall be anchored to masonry walls with $\frac{1}{2}$ in. (13 mm) diameter bolts 6 ft (1.8 m) on center or their equivalent. Bolts shall extend and be embedded at least 15 in. (381 mm) into the masonry, or be hooked or welded to not less than 0.20 in.^{2} (129 mm^2) of bond beam reinforcement placed not less than 6 in. (152 mm) from the top of the wall.

9.8.4 *Walls adjoining structural framing*—Where walls are dependent upon the structural frame for lateral support they shall be anchored to the structural members with metal anchors or otherwise keyed to the structural members. Metal anchors shall consist of $\frac{1}{2}$ in. (13 mm) bolts spaced at 4 ft (1.2 m) on center embedded 4 in. (102 mm) into the masonry, or their equivalent area.

9.9—Miscellaneous requirements

9.9.1 *Chases and recesses*—Masonry directly above chases or recesses wider than 12 in. (305 mm) shall be supported on lintels.

9.9.2 *Lintels*—The design for lintels shall be in

accordance with the provisions of Sections 5.6 and 7.3.4. Minimum end bearing shall be 4 in. (102 mm).

9.9.3 *Support on wood*—No masonry shall be supported on wood girders or other forms of wood construction.

9.9.4 *Corbelling*—Solid masonry units shall be used for corbelling. The maximum corbelled projection beyond the face of the wall shall be not more than one-half of the wall thickness or one-half the wythe thickness for hollow walls; the maximum projection of one unit shall neither exceed one-half the height of the unit nor one-third its thickness at right angles to the wall.

CHAPTER 10—SEISMIC DESIGN REQUIREMENTS

10.1—Scope

10.1.1 The seismic design requirements of this Chapter apply to the design of masonry and the construction of masonry building elements, except glass unit masonry and masonry veneers, for all seismic performance categories as defined in ASCE 7.

10.2—General

10.2.1 Masonry structures and masonry elements shall comply with the requirements of Sections 10.3 through 10.7 based on Seismic Performance Category A, B, C, D, or E as defined in ASCE 7. In addition, masonry structures and masonry elements shall comply with either the requirements of Section 5.1 or the requirements of Section 10.2.2.

10.2.2 *Strength requirements*—For masonry structures that are not designed in accordance with Section 5.1, the provisions of this section shall apply. The design strength of masonry structures and masonry elements shall be at least equal to the required strength determined in accordance with this section, except for masonry structures and masonry elements in Seismic Performance Category A designed in accordance with the provisions of Chapter 9.

10.2.2.1 *Required strength*—Required strength, U, to resist the seismic forces in such combinations with gravity and other loads, including load factors, shall be as required in the earthquake loads section of ASCE 7, except that nonbearing masonry walls shall be designed for the seismic force applied perpendicular to the plane of the wall and uniformly distributed over the wall area in lieu of the provisions of ASCE 7 Section 9.8.1.1.

10.2.2.2 *Nominal strength*—The nominal strength of masonry shall be taken as $2\frac{1}{2}$ times the allowable stress value. The allowable stress values shall be determined in accordance with Chapter 6 or 7 and are permitted to be increased by one-third for load combinations including earthquake.

10.2.2.3 *Design strength*—The design strength of masonry provided by a member, its connections to other members and its cross sections in terms of flexure, axial load, and shear shall be taken as the nominal strength multiplied by a strength reduction factor, ϕ.

(a) Axial load and flexure except for flexural tension in unreinforced masonry . . . . $\phi = 0.8$

(b) Flexural tension in unreinforced masonry . . . . $\phi = 0.4$

(c) Shear . . . . $\phi = 0.6$

(d) Shear and tension on anchor bolts embedded in masonry . . . . $\phi = 0.6$

10.2.2.4 *Drift limits*—The calculated story drift of masonry structures due to the combination of seismic forces and gravity loads shall not exceed 0.007 times the story height.

10.3—Seismic Performance Category A

10.3.1 Structures in Seismic Performance Category A shall comply with the requirements of Chapter 6, 7, or 9.

10.3.2 *Anchorage of masonry walls*—Masonry walls shall be anchored to the roof and all floors that provide lateral support for the wall. The anchorage shall provide a direct connection between the walls and the floor or roof construction. The connections shall be capable of resisting the greater of a seismic lateral force induced by the wall or 1000 times the effective peak velocity-related acceleration, lb per lineal ft of wall (14590 times, N/m).

10.4—Seismic Performance Category B

10.4.1 Structures in Seismic Performance Category B shall comply with the requirements of Seismic Performance Category A and to the additional requirements of this section.

10.4.2 The lateral force resisting system shall be designed to comply with the requirements of Chapter 6 or 7.

10.5—Seismic Performance Category C

10.5.1 Structures in Seismic Performance Category C shall comply with the requirements of Seismic Performance Category B and to the additional requirements of this section.

10.5.2 *Design of elements that are not part of lateral force-resisting system*

10.5.2.1 Load-bearing frames or columns that are not part of the lateral force resisting system shall be analyzed as to their effect on the response of the system. Such frames or columns shall be adequate for vertical load carrying capacity and induced moment due to the design story drift.

10.5.2.2 Masonry partition walls, masonry screen walls and other masonry elements that are not designed to resist vertical or lateral loads, other than those induced by their own mass, shall be isolated from the structure so that vertical and lateral forces are not imparted to these elements. Isolation joints and connectors between these elements and the structure shall be designed to accommodate the design story drift.

10.5.2.3 *Reinforcement requirements for masonry elements*—Masonry elements listed in Section 10.5.2.2 shall be reinforced in either the horizontal or vertical direction in accordance with the following:

(a) *Horizontal reinforcement*—Horizontal joint reinforcement shall consist of at least two longitudinal W1.7 wires spaced not more than 16 in.(406 mm) for walls greater than 4 in. (102 mm) in width and at least one longitudinal W1.7 wire spaced not more 16 in. (406 mm) for walls not exceeding 4 in. in width; or at least one No. 4 bar spaced not more than 48 in. (1219 mm). Where two longitudinal wires of joint reinforcement are used, the space between these wires shall be the widest that the mortar joint will accomodate. Horizontal reinforcement shall be provided within 16 in. (406 mm) of the top and bottom of these masonry elements.

(b) *Vertical reinforcement*—Vertical reinforcement shall consist of at least one No. 4 bar spaced not more than 48 in. (1219 mm). Vertical reinforcement shall be located within 16 in. (406 mm) of the ends of masonry walls.

10.5.3 *Design of elements that are part of the lateral force-resisting system*

10.5.3.1 *Connections to masonry shear walls*—Connectors shall be provided to transfer forces between masonry walls and horizontal elements in accordance with the requirements of Section 5.11. Connectors shall be designed to transfer horizontal design forces acting either perpendicular or parallel to the wall, but not less than 200 lb per lineal ft (2919 N/m) of wall. The maximum spacing between connectors shall be 4 ft (1.22 m).

10.5.3.2 *Connections to masonry columns*—Connectors shall be provided to transfer forces between masonry columns and horizontal elements in accordance with the requirements of Section 5.11. Where anchor bolts are used to connect horizontal elements to the tops of columns, anchor bolts shall be placed within lateral ties. Lateral ties shall enclose both the vertical bars in the column and the anchor bolts. There shall be a minimum of two No. 4 lateral ties provided in the top 5 in. (127 mm) of the column.

10.5.3.3 *Minimum reinforcement requirements for masonry shear walls*—Vertical reinforcement of at least 0.2 in.^{2} (129 mm^2) in cross-sectional area shall be provided at corners, within 16 in. (406 mm) of each side of openings, within 8 in. (203 mm) of each side of movement joints, within 8 in. (203 mm) of the ends of walls, and at a maximum spacing of 10 ft (3.1 m).

Horizontal joint reinforcement shall consist of at least two wires of W1.7 spaced not more than 16 in. (406 mm); or bond beam reinforcement shall be provided of at least 0.2 in.^{2} (129 mm^2) in cross-sectional area spaced not more than 10 ft (3.1 m). Horizontal reinforcement shall also be provided at the bottom and top of wall openings and shall extend not less than 24 in. (610 mm) nor less than 40 bar diameters past the opening; continuously at structurally connected roof and floor levels; and within 16 in. (406 mm) of the top of walls.

10.6—Seismic Performance Category D

10.6.1 Structures in Seismic Performance Category D shall comply with the requirements of Seismic Performance Category C and to the additional requirements of this section.

10.6.2 *Design requirements*—Masonry elements other than those covered by Section 10.5.2.2 shall be designed in accordance with the requirements of Chapter 7.

10.6.3 *Minimum reinforcement requirements for masonry walls*—Masonry walls other than those covered by Section 10.5.2.3 shall be reinforced in both the vertical and horizontal direction. The sum of the cross-sectional area of horizontal and vertical reinforcement shall be at least 0.002 times the gross cross-sectional area of the wall, and the minimum cross-sectional area in each direction shall be not less than 0.0007 times the gross cross-sectional area of the

wall. Reinforcement shall be uniformly distributed. The maximum spacing of reinforcement shall be 48 in. (1219 mm) provided that the walls are solid grouted and constructed of hollow open-end units, hollow units laid with full head joints or two wythes of solid units. The maximum spacing of reinforcement shall be 24 in. (610 mm) for all other masonry.

10.6.3.1 *Shear wall reinforcement requirements*—The maximum spacing of vertical and horizontal reinforcement shall be the smaller of; one-third the length of the shear wall, one-third the height of the shear wall, 48 in. (1219 mm). The minimum cross-sectional area of vertical reinforcement shall be one-third of the required shear reinforcement.

Shear reinforcement shall be anchored around vertical reinforcing bars with a standard hook.

10.6.4 *Minimum reinforcement for masonry columns*—Lateral ties in masonry columns shall be spaced not more than 8 in. (203 mm) on center and shall be at least $^3/_8$ in. (9.5 mm) diameter. Lateral ties shall be embedded in grout.

10.6.5 *Material requirements*—Neither Type N mortar nor masonry cement shall be used as part of the lateral force-resisting system.

10.6.6 *Lateral tie anchorage*—Standard hooks for lateral tie anchorage shall be either a 135 degree standard hook or a 180 degree standard hook.

10.7—Seismic Performance Category E

10.7.1 Structures in Seismic Performance Category E shall comply with the requirements of Seismic Performance Category D and to the additional requirements of this section.

10.7.2 *Design of elements that are not part of lateral force resisting system*—Stack bond masonry that is not part of the lateral force-resisting system shall have a horizontal cross-sectional area of reinforcement of at least 0.0015 times the gross cross-sectional area of masonry. The maximum spacing of horizontal reinforcement shall be 24 in. (610 mm). These elements shall be solidly grouted and shall be constructed of hollow open-end units or two wythes of solid units.

10.7.3 *Design of elements that are part of the lateral force resisting system*—Stack bond masonry that is part of the lateral force-resisting system shall have a horizontal cross-sectional area of reinforcement of at least 0.0025 times the gross cross-sectional area of masonry. The maximum spacing of horizontal reinforcement shall be 16 in. (406 mm). These elements shall be solidly grouted and shall be constructed of hollow open-end units or two wythes of solid units.

CHAPTER 11—GLASS UNIT MASONRY

11.1—Scope

11.1.1 This chapter covers empirical requirements for glass unit masonry elements in exterior or interior walls, either as isolated panels or in continuous bands.

11.2—Units

11.2.1 *General*—Hollow or solid glass block units shall be standard or thin units.

11.2.2 *Standard units*—The specified thickness shall be $3^7/_8$ in. (98 mm) thick.

11.2.3 *Thin units*—The specified thickness shall be $3^1/_8$ in. (79 mm) for hollow units or 3 in. (76 mm) for solid units.

11.3—Panel size

11.3.1 *Exterior standard-unit panels*—The maximum area of each individual standard-unit panel shall be 144 ft^2 (13.4 m^2) when the design wind pressure is 20 psf (958 Pa). The maximum panel dimension between structural supports shall be 25 ft (7.6 m) wide or 20 ft (6.1 m) high. Adjust the panel areas per Fig. 11.3-1 for other wind pressures.

11.3.2 *Exterior thin-unit panels*—The maximum area of each individual thin-unit panel shall be 85 ft^2 (7.9 m^2). The maximum dimension between structural supports shall be 15 ft (4.6 m) wide or 10 ft (3.1 m) high. Thin units shall not be used in applications where the design wind pressure exceeds 20 psf (958 Pa).

11.3.3 *Interior panels*—The maximum area of each individual standard-unit panel shall be 250 ft^2 (23.2 m^2). The maximum area of each thin-unit panel shall be 150 ft^2 (13.9 m^2). The maximum dimension between structural supports shall be 25 ft (7.6 m) wide or 20 ft (6.1 m) high.

11.3.4 *Curved panels*—The width of curved panels shall conform to the requirements of Sections 11.3.1, 11.3.2, and 11.3.3, except additional structural supports shall be provided at locations where a curved section joins a straight section, and at inflection points in multicurved walls.

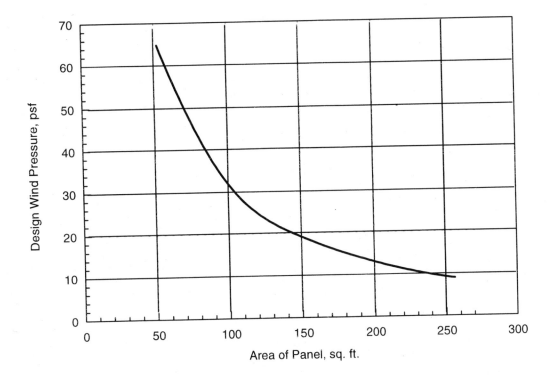

Fig. 11.3-1—Glass Unit Masonry Design Wind Load Resistance

11.4—Support

11.4.1 *Isolation*—Glass unit masonry panels shall be isolated so that in-plane loads are not imparted to the panel.

11.4.2 *Vertical*—Maximum total deflection of structural members supporting glass unit masonry shall not exceed $l/600$.

11.4.3 *Lateral*—Glass unit masonry panels shall be laterally supported along the top and sides of the panel. Lateral support shall be provided by panel anchors along the top and sides spaced not more than 16 in. (406 mm) on center or by channel-type restraints. Glass unit masonry panels shall be recessed at least 1 in. (25 mm) within channels and chases. Channel-type restraints must be oversized to accommodate expansion material in the opening, and packing and sealant between the framing restraints and the glass unit masonry perimeter units. Lateral supports for glass unit masonry panels shall be designed to resist applied loads, or a minimum of 200 lb per lineal ft (2918 N/m) of panel, whichever is greater.

11.5—Expansion joints

11.5.1 Glass unit masonry panels shall be provided with expansion joints along the top and sides at all structural supports. Expansion joints shall have sufficient thickness to accommodate displacements of the supporting structure, but shall not be less than $^3/_8$ in. (9.5 mm) in thickness. Expansion joints shall be entirely free of mortar or other debris and shall be filled with resilient material.

11.6—Mortar

11.6.1 Glass unit masonry shall be laid with Type S or N mortar.

11.7—Reinforcement

11.7.1 Glass unit masonry panels shall have horizontal joint reinforcement spaced not more than 16 in. (406 mm) on center, located in the mortar bed joint, and extending the entire length of the panel but not across expansion joints. Longitudinal wires shall be lapped a minimum of 6 in. (152 mm) at splices. Joint reinforcement shall be placed in the bed joint immediately below and above openings in the panel. The reinforcement shall have not less than two parallel longitudinal wires of size W1.7, and have welded cross wires of size W1.7.

CHAPTER 12—VENEERS

12.1—Scope

12.1.1 This chapter covers the requirements for design and detailing of anchored masonry veneer and its anchors. The veneer wythe is not subject to the allowable flexural tensile stress limitations of Chapter 6. Anchored veneer shall meet the requirements of Section 12.2 and shall be designed rationally by Section 12.3 or detailed by the prescriptive requirements of Sections 12.4 through 12.12.

12.1.2 All materials and construction shall comply with Section 3.1.1, except that Articles 1.4, 3.4D, and 3.6A of ACI 530.1/ASCE 6/TMS 602 shall not apply.

12.1.3 Adhered veneer is not covered under this Code. Any such veneer shall be considered a Special System and submitted accordingly to the Building Official.

12.2—General design requirements

12.2.1 Design and detail the backing system of exterior veneer to resist water penetration. Exterior sheathing shall be covered with a water-resistant membrane unless the sheathing is water resistant and the joints are sealed.

12.2.2 Design and detail flashing and weep holes in exterior veneer to divert water to the exterior. Weepholes shall be at least $^3/_{16}$ in. (5 mm) in diameter and spaced less than 33 in. (838 mm) on center.

12.2.3 Design and detail the veneer to accommodate differential movement.

12.2.4 Compliance with this Chapter exempts the requirements of Section 1.2.2(c). Thus, the inclusion of specified compressive strength of masonry in the Contract Documents is not required.

12.3—Alternative design of anchored masonry veneer

12.3.1 The design of anchored veneer is permitted under Section 1.3. The alternative design method shall have the following conditions:

(a) Loads shall be distributed through the veneer wythe to the anchors and the backing using principles of mechanics.

(b) Out of plane deflection of the backing shall be limited to maintain veneer stability.

(c) All masonry elements, other than the veneer wythe, shall meet the appropriate provisions of Chapters 1 through 6 or Chapters 1 through 5, 7 and 8, or Chapters 1 through 5 and 9.

(d) The veneer wythe is not subject to the provisions of Chapter 6.

(e) The provisions of Sections 12.1, 12.2, 12.11 and 12.12 shall apply.

12.4—Prescriptive requirements for anchored masonry veneer

12.4.1 Prescriptive requirements for anchored masonry veneer shall not be used in areas where the velocity pressure exceeds 25 lb/ft² (1197 Pa) as defined in ASCE 7.

12.4.2 Detail anchored veneer in compliance with Sections 12.4 through 12.12.

12.4.3 Connect anchored veneer to the backing with anchors that comply with Section 12.7 and Specification Article 2.4.

12.5—Vertical support of anchored masonry veneer

12.5.1 The weight of anchored veneer shall be supported vertically on concrete or masonry foundations or other noncombustible structural supports, except as permitted in Sections 12.5.1.1 and 12.5.1.4.

12.5.1.1 Anchored veneer is permitted to be supported vertically by preservative-treated wood foundations. The height of veneer supported by wood foundations shall not exceed 18 ft (5.5 m) above the support.

12.5.1.2 Anchored veneer with a backing of wood framing shall not exceed the height above the noncombustible foundation given in Table 12.5.1.

12.5.1.3 If anchored veneer with a backing of cold-formed steel framing exceeds the height above the noncombustible foundation given in Table 12.5.1, the weight of the veneer shall be supported by noncombustible construction for each story above the height limit given in Table 12.5.1.

12.5.1.4 When anchored veneer is used as an interior finish on wood framing, it shall have a weight of 40 lb/ft² (1915 Pa) or less and be installed in conformance with the provisions of this Chapter.

Table 12.5.1—Height limit from foundation

Height at plate, ft (m)	Height at gable, ft (m)
30 (9.1)	38 (11.6)

12.5.2 When anchored veneer is supported by floor construction, the floor shall be designed to limit deflection as required in Section 5.6.

12.5.3 Provide noncombustible lintels or supports attached to noncombustible framing over all openings where the anchored veneer is not self-supporting. The deflection of such lintels or supports shall conform to the requirements of Section 5.6.

12.6—Masonry units

12.6.1 Masonry units shall be at least $2^5/_8$ in. (67 mm) in actual thickness.

12.7—Anchor requirements

12.7.1 *Corrugated sheet metal anchors*

12.7.1.1 Corrugated sheet metal anchors shall be at least $^7/_8$ in. (22 mm) wide, have a base metal thickness of at least 0.03 in. (0.76 mm), and shall have corrugations with a wavelength of 0.3 to 0.5 in. (7.6 to 12.7 mm) and an amplitude of 0.06 to 0.10 in. (1.5 to 2.5 mm).

12.7.1.2 Corrugated sheet metal anchors shall be placed as follows:

(a) With solid units, embed anchors shall be embedded in the mortar joint and extend into the veneer wythe a minimum of $1^1/_2$ in. (38 mm), with at least $^5/_8$ in. (16 mm) mortar cover to the outside face.

(b) With hollow units, embed anchors in mortar or grout and extend into the veneer wythe a minimum of $1^1/_2$ in.

(38 mm), with at least $^5/_8$ in. (16 mm) mortar or grout cover to the outside face.

12.7.2 *Sheet metal anchors*

12.7.2.1 Sheet metal anchors shall be at least $^7/_8$ in. (22 mm) wide, have a base metal thickness of at least 0.06 in. (1.5 mm) and shall:

    (a) have corrugations as given in Section 12.7.1.1, or

    (b) be bent, notched, or punched to provide equivalent performance in pull-out or push-through.

12.7.2.2 Sheet metal anchors shall be placed as follows:

    (a) With solid units, embed anchors in the mortar joint and extend into the veneer wythe a minimum of $1^1/_2$ in. (38 mm), with at least $^5/_8$ in. (16 mm) mortar cover to the outside face.

    (b) With hollow units, embed anchors in mortar or grout and extend into the veneer wythe a minimum of $1^1/_2$ in. (38 mm), with at least $^5/_8$ in. (16 mm) mortar or grout cover to the outside face.

12.7.3 *Wire anchors*

12.7.3.1 Wire anchors shall be at least wire size W1.7 and have ends bent to form an extension from the bend at least 2 in. (51 mm) long.

12.7.3.2 Wire anchors shall be placed as follows:

    (a) With solid units, embed anchors in the mortar joint and extend into the veneer wythe a minimum of $1^1/_2$ in. (38 mm), with at least $^5/_8$ in. (16 mm) mortar cover to the outside face.

    (b) With hollow units, embed anchors in mortar or grout and extend into the veneer wythe a minimum of $1^1/_2$ in. (38 mm), with at least $^5/_8$ in. (16 mm) mortar or grout cover to the outside face.

12.7.4 *Joint reinforcement*

12.7.4.1 Ladder-type or tab-type joint reinforcement is permitted. Cross wires used to anchor masonry veneer shall be at least wire size W1.7 and shall be spaced at a maximum of 16 in. (406 mm) on center. Cross wires shall be welded to longitudinal wires, which shall be at least wire size W1.7.

12.7.4.2 Embed longitudinal wires of joint reinforcement in the mortar joint with at least $^5/_8$ in. (16 mm) mortar cover on each side.

12.7.5 *Adjustable anchors*

12.7.5.1 Sheet metal and wire components of adjustable anchors shall conform to the requirements of Sections 12.7.1, 12.7.2 or 12.7.3. Adjustable anchors with joint reinforcement shall also meet the requirements of Section 12.7.4.

12.7.5.2 Maximum clearance between connecting parts of the tie shall be $^1/_{16}$ in. (2 mm).

12.7.5.3 Adjustable anchors shall be detailed to prevent disengagement.

12.7.5.4 Pintle anchors shall have at least two pintle legs of wire size W2.8 each and shall have an offset not exceeding $1^1/_4$ in. (32 mm).

12.7.5.5 Adjustable anchors of equivalent strength and stiffness to those specified in Sections 12.7.5.1 through 12.7.5.4 are permitted.

12.7.6 *Anchor spacing*

12.7.6.1 For adjustable two-piece anchors, anchors of wire size W1.7, and 22 gage (0.76 mm) corrugated sheet metal anchors, provide at least one anchor for each 2.67 ft^2 (0.25 m^2) of wall area.

12.7.6.2 For all other anchors, provide at least one anchor for each 3.5 ft^2 (0.33 m^2) of wall area.

12.7.6.3 Space anchors at a maximum of 32 in. (813 mm) horizontally and 18 in. (457 mm) vertically.

12.7.6.4 Provide additional anchors around all openings larger than 16 in. (406 mm) in either dimension. Space anchors around perimeter of opening at a maximum of 3 ft (1 m) on center. Place anchors within 12 in. (305 mm) of openings.

12.7.7 *Joint thickness for anchors*—Mortar bed joint thickness shall be at least twice the thickness of the embedded anchor.

12.8—Masonry veneer anchored to wood backing

12.8.1 Veneer shall be attached with any anchor permitted in Section 12.7.

12.8.2 Attach each anchor to wood studs or wood framing with a corrosion-resistant 8d common nail, or fastener with equivalent or greater pullout strength. For corrugated sheet metal anchors, locate the nail or fastener within $^1/_2$ in. (13 mm) of the 90 deg bend in the anchor.

12.8.3 Maintain a maximum distance between the inside face of the veneer and outside face of the solid sheathing of 1 in. (25 mm) when corrugated anchors are used. Maintain a maximum distance between the inside face of the veneer and the wood stud or wood framing of $4^1/_2$ in. (114 mm) when other anchors are used. Maintain a 1 in. (25 mm) minimum air space.

12.9—Masonry veneer anchored to steel backing

12.9.1 Attach veneer with adjustable anchors.

12.9.2 Attach each anchor to steel framing with corrosion-resistant screws that have a minimum nominal shank diameter of 0.190 in. (4.8 mm).

12.9.3 Cold-formed steel framing shall be corrosion resistant and have a minimum base metal thickness of 0.043 in. (1.1 mm).

12.9.4 Maintain a $4^1/_2$ in. (114 mm) maximum distance between the inside face of the veneer and the steel framing. Maintain a 1 in. (25 mm) minimum air space.

12.10—Masonry veneer anchored to masonry or concrete backing

12.10.1 Attach veneer to masonry backing with wire anchors, adjustable anchors, or joint reinforcement. Attach veneer to concrete backing with adjustable anchors.

12.10.2 Maintain a $4^1/_2$ in. (114 mm) maximum distance between the inside face of the veneer and the outside face of the masonry or concrete backing. Maintain a 1 in. (25 mm) minimum air space.

12.11—Veneer laid in other than running bond

12.11.1 Anchored veneer laid in other than running bond shall have joint reinforcement of at least one wire, of size W1.7, spaced at a maximum of 18 in. (457 mm) on center vertically.

12.12—Requirements in seismic areas

12.12.1 *Seismic Performance Category C*

12.12.1.1   The requirements of this section apply to anchored veneer for buildings in Seismic Performance Category C.

12.12.1.2   Isolate the sides and top of anchored veneer from the structure so that vertical and lateral seismic forces resisted by the structure are not imparted to the veneer wythe.

12.12.2 *Seismic Performance Category D*

12.12.2.1   The requirements for Seismic Performance Category C and the requirements of this section apply to anchored veneer for buildings in Seismic Performance Category D.

12.12.2.2   Support the weight of anchored veneer for each story independent of other stories.

12.12.2.3   Reduce the maximum wall area supported by each anchor to 75 percent of that required in Sections 12.7.6.1 and 12.7.6.2.   Maximum horizontal and vertical spacings are unchanged.

12.12.2.4   Provide continuous, single-wire joint reinforcement of minimum wire size W1.7 at a maximum spacing of 18 in. (457 mm) on center vertically.

12.12.3 *Seismic Performance Category E*

12.12.3.1   The requirements for Seismic Performance Category D and the requirements of this section apply to anchored veneer for buildings in Seismic Performance Category E.

12.12.3.2   Provide vertical expansion joints at all returns and corners.

12.12.3.3   Mechanically attach anchors to the joint reinforcement required in Section 12.12.2.4 with clips or hooks.

TRANSLATION OF INCH-POUND UNITS TO SI UNITS

FOR INFORMATION ONLY

Code Equations

Eq. No. or Sec. No.	Equation	Units
5.8.2.2(c)	$0.083 \sqrt{unit\ compressive\ strength\ of\ header}$ (over net area of header)	$\sqrt{unit\ compressive\ strength\ of\ header}$ in MPa
(5-1)	$B_a = 0.042 A_p \sqrt{f'_m}$	A_p in mm^2 B_a in Newtons $\sqrt{f'_m}$ in MPa
(5-2)	$B_a = 0.2 A_b f_y$	A_b in mm^2 B_a in Newtons f_y in MPa
(5-3)	$A_p = \pi l_b^2$	A_p in mm^2 l_b in mm
(5-4)	$A_p = \pi l_{be}^2$	A_p in mm^2 l_{be} in mm
(5-5)	$B_v = 1070 \sqrt[4]{f'_m A_b}$	A_b in mm^2 B_v in Newtons $\sqrt{f'_m}$ in MPa
(5-6)	$B_v = 0.12 A_b f_y$	A_b in mm^2 B_v in Newtons f_y in MPa
(5-7)	$\dfrac{b_a}{B_a} + \dfrac{b_v}{B_v} \leq 1$	B_a in Newtons B_v in Newtons b_a in Newtons b_v in Newtons
(6-1)	$\dfrac{f_a}{F_a} + \dfrac{f_b}{F_b} \leq 1$	F_a in MPa F_b in MPa f_a in MPa f_b in MPa
(6-2)	$P \leq (^1/_4) P_e$	P in Newtons P_e in Newtons
(6-3)	$F_a = (^1/_4) f'_m \left[1 - \left(\dfrac{h}{140r} \right)^2 \right]$	F_a in MPa f'_m in MPa h in MPa r in MPa
(6-4)	$F_a = (^1/_4) f'_m \left(\dfrac{70r}{h} \right)^2$	F_a in MPa f'_m in MPa h in mm r in mm
(6-5)	$F_b = (^1/_3) f'_m$	F_b in MPa f'_m in MPa

Eq. No. or Sec. No.	Equation	Units
(6-6)	$$P_e = \frac{\pi^2 E_m I}{h^2}\left(1 - 0.577\,\frac{e}{r}\right)^3$$	E_m in MPa e in mm h in mm I in mm^4 P_e in MPa r in mm
(6-7)	$$f_v = \frac{VQ}{Ib}$$	b in mm f_v in MPa I in mm^4 Q in mm^3 V in Newtons
6.5.2(a)	$$0.125\,\sqrt{f'_m}$$	$\sqrt{f'_m}$ in MPa
6.5.2(c)	$$v + 0.45\,N_v/A_n$$	A_n in mm^2 N_v in Newtons v in MPa Answer in MPa
(7-1)	$$P_a = (0.25\,f'_m A_n + 0.65\,A_{st}F_s)\left[1 - \left(\frac{h}{140\,r}\right)^2\right]$$	A_n in mm^2 h in mm A_{st} in mm^2 P_a in Newtons F_s in MPa r in mm f'_m in MPa
(7-2)	$$P_a = (0.25\,f_m'A_n + 0.65\,A_{st}F_s)\left(\frac{70r}{h}\right)^2$$	A_n in mm^2 h in mm A_{st} in mm^2 P_a in Newtons F_s in MPa r in mm f'_m in MPa
(7-3)	$$f_v = \frac{V}{bd}$$	b in mm d in mm f_v in MPa V in Newtons
(7-4)	$$F_v = 0.083\,\sqrt{f'_m}$$	F_v in MPa $\sqrt{f'_m}$ in MPa
(7-5)	$$F_v = 0.028\,[4 - (M/Vd)]\,\sqrt{f'_m}$$ but shall not exceed $$0.55 - 0.31\,(M/Vd)\ \text{in MPa}$$	d in mm F_v in MPa M in Newton•mm V in Newtons $\sqrt{f'_m}$ in MPa
(7-6)	$$F_v = 0.083\,\sqrt{f'_m}$$	F_v in MPa $\sqrt{f'_m}$ in MPa
(7-7)	$$F_v = 0.25\,\sqrt{f'_m}$$	F_v in MPa $\sqrt{f'_m}$ in MPa

Eq. No. or Sec. No.	Equation	Units
(7-8)	$F_v = 0.042 [4 - (M/Vd)] \sqrt{f'_m}$ but shall not exceed $0.82 - 0.31 (M/Vd)$ in MPa	d in mm F_v in MPa M in Newton•mm V in Newtons $\sqrt{f'_m}$ in MPa
(7-9)	$F_v = 0.125 \sqrt{f'_m}$	F_v in MPa $\sqrt{f'_m}$ in MPa
(7-10)	$A_v = \dfrac{Vs}{F_s d}$	A_v in mm^2 d in mm F_s in MPa s in mm V in Newtons
(8-1)	$l_d = 0.22 d_b F_s$	d_b in mm F_s in MPa l_d in mm
(8-2)	$l_d = 0.29 d_b F_s$	d_b in mm F_s in MPa l_d in mm

Specification for Masonry Structures
(ACI 530.1-95/ASCE 6-95/TMS 602-95)

Reported by the Masonry Standards Joint Committee

James Colville
Chairman

Max L. Porter
Vice Chairman

J. Gregg Borchelt
Secretary

Maribeth S. Bradfield
Membership Secretary

Regular Members[1]:

Gene C. Abbate
Bechara E. Abboud
Bijan Ahmadi
Amde M. Amde
Richard H. Atkinson
William G. Bailey
Stuart R. Beavers
Robert J. Beiner
Frank Berg
Russell H. Brown
A. Dwayne Bryant
Kevin D. Callahan
Mario J. Catani
Robert W. Crooks
Kenneth G. Dagostino, Jr.

Gerald A. Dalrymple
Steve Dill
Russell T. Flynn
John A. Frauenhoffer
Thomas A. Gangel
Richard M. Gensert
Satyendra K. Ghosh
Clayford T. Grimm
John C. Grogan
Craig K. Haney
Gary C. Hart
Barbara Heller
Robert Hendershot
Mark B. Hogan
Thomas A. Holm

Rochelle C. Jaffe
John C. Kariotis
Richard E. Klingner
Walter Laska
L. Donald Leinweber
Hugh C. MacDonald, Jr.
Billy R. Manning
John H. Matthys
Robert McCluer
Donald G. McMican
George A. Miller
Reg Miller
Colin C. Munro
W. Thomas Munsell
Antonio Nanni

Joseph F. Neussendorfer
Joseph E. Saliba
Arturo Schultz
Matthew J. Scolforo
Daniel Shapiro
John M. Sheehan
Robert A. Speed
Ervell A. Staab
Jerry G. Stockbridge
Itzhak Tepper
Robert C. Thacker
Donald W. Vannoy
Terence A. Weigel
A. Rhett Whitlock

Associate Members[2]:

James E. Amrhein
David T. Biggs
James W. Cowie
John Chrysler
Terry M. Curtis
Walter L. Dickey
Jeffrey L. Elder
Brent A. Gabby

Hans R. Ganz
H. R. Hamilton, III
B. A. Haseltine
Edwin G. Hedstrom
A. W. Hendry
Thomas F. Herrell
Steve Lawrence
Nicholas T. Loomis

Robert F. Mast
John Melander
Raul Alamo Neihart
Robert L. Nelson
Rick Okawa
Adrian W. Page
Ruiz Lopez M. Rafael
Roscoe Reeves, Jr.

Phillip J. Samblanet
Richard C. Schumacher
John G. Tawresey
Robert D. Thomas
Dean J. Tills
Charles W. C. Yancey

SYNOPSIS

This Specification for Masonry Structures (ACI 530.1-95/ASCE 6-95/TMS 602-95) is written as a master specification and is required by the Code to control materials, labor and construction. Thus, this Specification covers minimum construction requirements for masonry in structures. Included are quality assurance requirements for materials; the placing, bonding and anchoring of masonry; and the placement of grout and of reinforcement. This Specification is meant to be modified and referenced in the Project Manual.

Keywords: clay brick; clay tile; concrete block; concrete brick; construction; construction materials; curing; grout; grouting; inspection; joints; masonry; materials handling; mortars (material and placement); quality assurance and quality control; reinforcing steel; specifications; tests; tolerances.

---

[1] Regular members fully participate in Committee activities, including responding to correspondence and voting.

[2] Associate members monitor Committee activities, but do not have voting privileges.

Adopted as a standard of the American Concrete Institute November 1988, in accordance with the Institute's standardization procedure. Revised by the Institute's Expedited Standardization Procedure effective September 1, 1995. Adopted as a standard of the American Society of Civil Engineers August 1989, in accordance with the Society's standardization procedure and revised by the Society's standardization procedure effective August 1, 1995. Adopted as a standard of The Masonry Society, July 1, 1992, in accordance with the Society's standardization procedure and revised by the Society's standardization procedure and revised by the Society's standardization procedure effective August 1, 1995.

FOREWORD

F1. This foreword is included for explanatory purposes only; it does not form a part of Specification ACI 530.1/ASCE 6/TMS 602.

F2. Specification ACI 530.1/ASCE 6/TMS 602 is a Reference Standard which the Architect/ Engineer may cite in the Contract Documents for any project, together with supplementary requirements for the specific project.

F3. Specification ACI 530.1/ASCE 6/TMS 602 is written in the three-part section format of the Construction Specifications Institute, as adapted by ACI. The language is generally imperative and terse.

F4. A Specification Checklist is included as a preface to, but not forming a part of, Specification ACI 530.1/ASCE 6/TMS 602. The purpose of this Specification Checklist is to assist the Architect/Engineer in properly choosing and specifying the necessary mandatory and optional requirements for the Project Specifications.

PREFACE TO SPECIFICATION CHECKLIST

P1. Specification ACI 530.1/ASCE 6/TMS 602 is intended to be used in its entirety by reference in the Project Specifications. Individual sections, articles, or paragraphs should not be copied into the Project Specifications since taking them out of context may change their meaning.

P2. Building codes (of which this standard is a part by reference) set minimum requirements necessary to protect the public. Project Specifications often stipulate requirements more restrictive than the minimum. Adjustments to the needs of a particular project are intended to be made by the Architect/Engineer by reviewing each of the items in the Specification Checklist and then including the Architect/Engineer's decision on each item as a mandatory requirement in the Project Specifications.

P3. These mandatory requirements should designate the specific qualities, procedures, materials, and performance criteria for which alternatives are permitted or for which provisions were not made in the Specifications. Exceptions to this Specification should be made in the Project Specifications, if required.

P4. A statement such as the following will serve to make Specification ACI 530.1/ASCE 6/TMS 602 an official part of the Project Specifications:

Masonry construction and materials shall conform to all requirements of "Specification for Masonry Structures (ACI 530.1/ASCE 6/TMS 602)," published by the American Concrete Institute, Detroit, Michigan, except as modified by the requirements of these Contract Documents.

P5. The Specification Checklist which follows is addressed to each item of this Specification where the Architect/Engineer must or may make a choice of alternatives; may add provisions if not indicated; or may take exceptions. The Specification Checklist consists of two columns; the first identifies the sections, parts, and articles of the Specification, and the second column contains notes to the Architect/Engineer to indicate the type of action required by the Architect/Engineer.

MANDATORY SPECIFICATION CHECKLIST

Section/Part/Article		Notes to the Architect/Engineer
PART 1—GENERAL		
1.1B		Indicate the articles (parts and sections) of ACI 530.1/ASCE 6/TMS 602 excluded from the Project Specifications. List the articles of ACI 530.1/ASCE 6/TMS 602 at variance with the Project Specifications.
1.4A	Compressive strength requirements	Specify f'_m, except for veneer.
1.6	Quality assurance	Designate the Owner's representatives for quality assurance.
1.6A1		Specify quality assurance requirements.
PART 2—PRODUCTS		
2.1	Mortar type and materials	Specify type, color and cementitious materials to be used in mortar and mortar to be used for the various parts of the project and the type of mortar to be used with each type of masonry unit.
2.3	Masonry units	Specify the masonry units to be used for the various parts of the projects.
2.4	Reinforcement and metal accessories	Specify type and grade of reinforcement, connectors and accessories.
2.4B3	Welded wire fabric	Specify when welded wire fabric is to be plain.
2.4D	Stainless steel	Specify when stainless steel joint reinforcement, anchors, ties, and/or accessories are required.
2.4E	Coating for corrosion protection	Specify which interior walls are governed by this provision.
2.5D	Joint fillers	Specify size and shape of joint fillers.
2.7B	Prefabricated masonry	Specify prefabricated masonry and requirements in supplement of those of ASTM C 901.
PART 3—EXECUTION		
3.3D2	Pipes and conduits	Specify sleeve sizes and spacing.
3.3D6	Accessories	Specify accessories not indicated on the Project Drawings.
3.3D7	Movement joints	Indicate type and location of movement joints on the Project Drawings.

OPTIONAL SPECIFICATION CHECKLIST

Section/Part/Article	Notes to the Architect/Engineer
PART 1—GENERAL	
1.4B1 Determination of compressive strength	Specify method used to determine compressive strength.
1.4B2 Unit strength method	Specify when strength of grout is to be determined by test.
1.5B	Specify when required.
1.6A Testing laboratory services	Specify requirements at variance with ACI 530.1/ASCE 6/TMS 602.
1.6A4 Inspection and testing (a through f)	Specify testing and inspection requirements.
1.6B1	Specify sample panels needed.
1.6B2	Specify whether sample panels shall not be part of finished construction.
PART 2—PRODUCTS	
2.2A	Specify grout requirements at variance with ACI 530.1/ASCE 6/TMS 602. Specify admixtures.
2.5A and 2.5B Movement joint	Specify requirements at variance with ACI 530.1/ASCE 6/TMS 602.
2.5C2 Masonry cleaner	Specify where other than potable water, detergents, acid or caustic solutions are allowed and how to neutralize them.
2.6A Mortar	Specify if hand mixing is allowed and the method of measurement of material.
2.6A2 Pigment	Specify pigments.
2.6B1 Grout proportioning and mixing	Specify requirements at variance with ACI 530.1/ASCE 6/TMS 602.
2.6B2 Grout slump	Indicate when grout slump tests are required to verify slump.
PART 3—EXECUTION	
3.2C Wetting masonry units	Specify when units are to be wetted.
3.3A Bond pattern	Specify bond pattern other than running bond.
3.3B1 Bed and head joints	Specify thickness and tooling differing from ACI 530.1/ ASCE 6/TMS 602.
3.3B2 Collar joints	Specify the filling of collar joints less than $3/4$ in. (19 mm) thick differing from ACI 530.1/ASCE 6/TMS 602.
3.3B3 Hollow units	Specify when cross webs are to be mortar bedded.
3.3B4 Solid units	Specify mortar bedding at variance with ACI 530.1/ASCE 6/TMS 602.
3.3B5 Glass units	Specify mortar bedding at variance with ACI 530.1/ASCE 6/TMS 602.
3.3D2 Embedded items	Specify locations where sleeves are required for pipes or conduits.
3.3D8 Insulation	Specify insulation.
3.4D2, 3 and 4	Specify requirements at variance with ACI 530.1/ASCE 6/TMS 602.
3.6A Quality assurance prism test	See Article 1.4B1
3.6B Quality assurance mortar test	Specify mortar tests.

OPTIONAL SPECIFICATION CHECKLIST

Section/Part/Article	Notes to the Architect/Engineer
PART 3—EXECUTION 3.6C Quality assurance grout test	Specify if testing of grout is required.

SUBMITTALS

Section/Part/Article	Notes to the Architect/Engineer
1.5 Submittals	Specify submittals required under Article 1.5B.

CONTENTS

PART 1—GENERAL

1.1—Summary

A. This Specification covers requirements for materials and construction of masonry structures. Metric values shown in parentheses are provided for information only and are not part of this Specification.

B. The provisions of this Specification govern except where different provisions are specified.

C. Furnish and construct masonry in accordance with the requirements of the Contract Documents. This article covers the furnishing and construction of masonry including the following:

1. Furnishing and placing masonry units, grout, mortar, masonry lintels, sills, copings, through-wall flashing, and connectors.

2. Furnishing, erecting and maintaining of bracing, forming, scaffolding, rigging, and shoring.

3. Furnishing and installing other equipment for constructing masonry.

4. Cleaning masonry and removing surplus material and waste.

5. Installing lintels, nailing blocks, inserts, window and door frames, connectors, and construction items to be built into the masonry, and building in vent pipes, conduits and other items furnished and located by other trades.

1.2—Definitions

A. *Acceptable, accepted*—Acceptable to or accepted by the Architect/Engineer.

B. *Architect/Engineer*—The architect, engineer, architectural firm, engineering firm, or architectural and engineering firm, issuing Drawings and Specifications, or administering the Work under Contract Specifications and Project Drawings, or both.

C. *Area, gross cross-sectional*—The area delineated by the out-to-out dimensions of masonry in the plane under consideration.

D. *Area, net cross-sectional*—The area of masonry units, grout, and mortar crossed by the plane under consideration based on out-to-out dimensions.

E. *Collar joint*—Vertical, longitudinal joint between wythes of masonry or between masonry and back up construction which is permitted to be filled with mortar or grout.

F. *Compressive strength of masonry*—Maximum compressive force resisted per unit of net cross-sectional area of masonry, determined by the testing of masonry prisms; or a function of individual masonry units, mortar and grout in accordance with the provisions of this Specification.

G. *Contract Documents*—Documents establishing the required Work, and including in particular, the Project Drawings and Project Specifications.

H. *Contractor*—The person, firm, or corporation with whom the Owner enters into an agreement for construction of the Work.

I. *Dimension, nominal*—A nominal dimension is equal to a specified dimension plus an allowance of the joints with which the units are to be laid. Nominal dimensions are usually stated in whole numbers. Thickness is given first, followed by height and then length.

J. *Dimensions, specified*—Dimensions specified for the manufacture or construction of a unit, joint, or element.

K. *Glass unit masonry*—Nonload-bearing masonry composed of glass units bonded by mortar.

L. *Grout lift*—An increment of grout height within a total grout pour. A grout pour consists of one or more grout lifts.

M. *Grout pour*—The total height of masonry to be grouted prior to erection of additional masonry. A grout pour consists of one or more grout lifts.

N. *Mean daily temperature*—The average daily temperature of temperature extremes predicted by a local weather bureau for the next 24 hours.

O. *Minimum/maximum (not less than ... not more than)*—Minimum or maximum values given in this Specification are absolute. Do not construe that tolerances allow lowering a minimum or increasing a maximum.

P. *Otherwise required*—Specified differently in requirements supplemental to this Specification.

Q. *Owner*—The public body or authority, corporation, association, partnership, or individual for whom the Work is provided.

R. *Partition wall*—An interior wall without structural function.

S. *Project Drawings*—The Drawings which, along with the Project Specifications, complete the descriptive information for constructing the Work required or referred to in the Contract Documents.

T. *Project Specifications*—The written documents which specify requirements for a project in accordance with the service parameters and other specific criteria established by the Owner or his agent.

U. *Running bond*—The placement of masonry units such that head joints in successive courses are horizontally offset at least one-quarter the unit length.

V. *Specified compressive strength of masonry, f'_m*—Minimum compressive strength, expressed as force per unit of net cross-sectional area, required of the masonry used in construction by the project documents and upon which the project design is based.

W. *Stack bond*—For the purpose of this Specification, stack bond is other than running bond. Usually the placement of masonry units is such that head joints in successive courses are vertically aligned.

X. *Stone masonry*—Masonry composed of field, quarried, or cast stone units bonded by mortar.

1. *Stone masonry, ashlar*—Stone masonry composed of rectangular units having sawed, dressed, or squared bed surfaces and bonded by mortar.

2. *Stone masonry, rubble*—Stone masonry composed of irregular shaped units bonded by mortar.

Y. *Submit, submitted*—Submit, submitted to the Architect/Engineer for review.

Z. *Wall*—A vertical element with a horizontal length to thickness ratio greater than 3, used to enclose space.

AA. *Wall, load-bearing*—A wall carrying vertical loads greater than 200 lb per lineal foot (2918 N/m) in addition to its own weight.

AB. *Wall, masonry bonded hollow*—A multiwythe wall built with masonry units arranged to provide an air space between the wythes and with the wythes bonded together with masonry units.

AC. *When required*—Specified in requirements supplemental to this Specification.

AD. *Work*—The furnishing and performance of all equipment, services, labor, and materials required by the Contract Documents for the construction of masonry for the project or part of project under consideration.

AE. *Wythe*—Each continuous vertical section of a wall, one masonry unit in thickness.

1.3—References

Standards of the American Concrete Institute, the American Society for Testing and Materials, and the American Welding Society referred to in this Specification are listed below with their serial designations, including year of adoption or revision, and are declared to be part of this Specification as if fully set forth in this document except as modified herein.

A.	ACI 117-90	Standard Specifications for Tolerances for Concrete Construction and Materials
B.	ACI 315-92	Details and Detailing of Concrete Reinforcement
C.	ASTM A 36-90/ A 36M-93a	Specification for Structural Steel
D.	ASTM A 82-90a	Specification for Steel Wire, Plain, for Concrete Reinforcement
E.	ASTM A 123-89a^{e1}	Specifications for Zinc (Hot-Dip Galvanized) Coatings on Iron and Steel Products
F.	ASTM A 153-82	Specification for Zinc(1987) Coating (Hot-Dip) on Iron and Steel Hardware
G.	ASTM A 167-93	Specification for Stainless and Heat-Resisting Chromium-Nickel Steel Plate, Sheet, and Strip
H.	ASTM A 185-90a	Specification for Steel Welded Wire, Fabric, Plain, for Concrete Reinforcement
I.	ASTM A 307-93a	Specification for Carbon Steel Bolts and Studs, 60,000 psi Tensile Strength
J.	ASTM A 366-85/ A 366M-91	Specification for Steel, Sheet, Carbon, Cold-Rolled, Commercial Quality
K.	ASTM A 496-93	Specification for Steel Wire, Deformed, for Concrete Reinforcement
L.	ASTM A 497-90b	Specification for Welded Deformed Steel Wire Fabric for Concrete Reinforcement
M.	ASTM A 525-93	Specification for General Requirements for Steel Sheet, Zinc-Coated (Galvanized) by the Hot-Dip Process
N.	ASTM A 580-83	Specification for Stainless and Heat-Resisting Steel Wire
O.	ASTM A 615-90/	Specification for A 615M-93 Deformed and Plain Billet-Steel Bars for Concrete Reinforcement
P.	ASTM A 616-90/ A 616M-93	Specification for Rail-Steel Deformed and Plain Bars for Concrete Reinforcement
Q.	ASTM A 617-90/ A 617M-93	Specification for Axle-Steel Deformed and Plain Bars for Concrete Reinforcement
R.	ASTM A 641-92	Specification for Zinc-Coated (Galvanized) Carbon Steel Wire
S.	ASTM A 666-87	Specification for Austenitic Stainless Steel, Sheet, Strip, Plate and Flat Bar for Structural Applications
T.	ASTM A 706-90/ A 706M-93a	Specification for Low-Alloy Steel Deformed Bars for Concrete Reinforcement
U.	ASTM A 767/ A 767M-90	Specification for Zinc-Coated (Galvanized) Bars for Concrete Reinforcement
V.	ASTM A 775/ A 775M-92	Specification for Epoxy-Coated Reinforcing Steel Bars
W.	ASTM C 34-93	Specification for Structural Clay Load-Bearing Wall Tile
X.	ASTM C 55-93a	Specification for Concrete Building Brick
Y.	ASTM C 56-93	Specification for Structural Clay Non-Load-Bearing Tile
Z.	ASTM C 62-92c	Specification for Building Brick (Solid Masonry Units Made from Clay or Shale)
AA.	ASTM C 67-93a	Test Methods of Sampling and Testing Brick and Structural Clay Tile
AB.	ASTM C 73-85 (1989)	Specification for Calcium Silicate Face Brick (Sand-Lime Brick)
AC.	ASTM C 90-93	Specification for Load-Bearing Concrete Masonry Units

AD. ASTM C 97-90 — Test Methods for Absorption and Bulk Specific Gravity of Dimension Stone

AE. ASTM C 99-87 (1993) — Test Method for Modulus of Rupture of Dimension Stone

AF. ASTM C 120-90 — Method for Flexure Testing of Slate (Modulus of Rupture, Modulus of Elasticity)

AG. ASTM C 121-90 — Test Method for Water Absorption of Slate

AH. ASTM C 126-91 — Specification for Ceramic Glazed Structural Clay Facing Tile, Facing Brick, and Solid Masonry Units

AI. ASTM C 129-93 — Specification for Non-Load-Bearing Concrete Masonry Units

AJ. ASTM C 140-91 — Methods of Sampling and Testing Concrete Masonry Units

AK. ASTM C 143-90a — Test Method for Slump of Hydraulic Cement Concrete

AL. ASTM C 170-90 — Test Method for Compressive Strength of Dimension Stone

AM. ASTM C 212-93 — Specification for Structural Clay Facing Tile

AN. ASTM C 216-92d — Specification for Facing Brick (Solid Masonry Units Made from Clay or Shale)

AO. ASTM C 270-92a — Specification for Mortar for Unit Masonry

AP. ASTM C 476-91 — Specification for Grout for Masonry

AQ. ASTM C 503-89 — Specification for Marble Dimension Stone (Exterior)

AR. ASTM C 568-89$^{\epsilon 1}$ — Specification for Limestone Dimension Stone

AS. ASTM C 615-92 — Specification for Granite Dimension Stone

AT. ASTM C 616-89$^{\epsilon 1}$ — Specification for Quartz-Based Dimension Stone

AU. ASTM C 629-89$^{\epsilon 1}$ — Specification for Slate Dimension Stone

AV. ASTM C 652-93 — Specification for Hollow Brick (Hollow Masonry Units Made from Clay or Shale)

AW. ASTM C 744-92 — Specification for Prefaced Concrete and Calcium Silicate Masonry Units

AX. ASTM C 780-91 — Test Method for Preconstruction and Construction Evaluation of Mortars for Plain and Reinforced Unit Masonry

AY. ASTM C 901-93a — Specification for Prefabricated Masonry Panels

AZ. ASTM C 920-87 — Specification for Elastomeric Joint Sealants

BA. ASTM C 1019-89a$^{\epsilon 1}$ — Test Method for Sampling and Testing Grout

BB. ASTM D 994-71 (1982)$^{\epsilon 1}$ — Specification for Preformed Expansion Joint Filler for Concrete (Bituminous Type)

BC. ASTM D 1056-91 — Specification for Flexible Cellular Materials—Sponge or Expanded Rubber

BD. ASTM D 2000-90 — Classification System for Rubber Products in Automotive Applications

BE. ASTM D 2287-92 — Specification for Nonrigid Vinyl Chloride Polymer and Copolymer Molding and Extrusion Compounds

BF. ASTM E 447-92b — Test Methods for Compressive Strength of Masonry Prisms

BG. ANSI/AWS D1.4-92 — Structural Welding Code—Reinforcing Steel

1.4—System description

A. *Compressive strength requirements*—Compressive strength of masonry in each masonry wythe and grouted collar joint shall equal or exceed the applicable f'_m.

B. *Compressive strength determination*

1. *Alternatives for determination of compressive strength*—Determine the compressive strength for each wythe by the unit strength method or by the prism test method as specified herein.

2. *Unit strength method*

a. *Clay masonry*—Determine the compressive strength of masonry based on the strength of the units and the type of mortar specified using Table 1. The following Articles must be met:

1) Units conform to ASTM C 62, ASTM C 216, or ASTM C 652 and are sampled and tested in accordance with ASTM C 67.

2) Thickness of bed joints does not exceed $^5/_8$ in. (16 mm).

3) For grouted masonry, the grout meets one of the following requirements:

a) Grout conforms to ASTM C 476.

b) Grout compressive strength equals f'_m but compressive strength is not less than 2000 psi (13.8 MPa). Determine compressive strength of grout in accordance with ASTM C 1019.

Table 1—Compressive strength of masonry based on the compressive strength of clay masonry units and type of mortar used in construction

Net area compressive strength of clay masonry units, psi (MPa)		Net area compressive strength of masonry, psi (MPa)
Type M or S mortar	Type N mortar	
2400 (16.5)	3000 (20.1)	1000 (6.9)
4400 (30.3)	5500 (37.9)	1500 (10.3)
6400 (44.1)	8000 (55.1)	2000 (13.8)
8400 (57.9)	10,500 (72.4)	2500 (17.2)
10,400 (71.7)	13,000 (89.6)	3000 (20.1)
12,400 (85.4)	—	3500 (24.1)
14,400 (99.2)	—	4000 (27.6)

b. *Concrete masonry*—Determine the compressive strength of masonry, based on the strength of the unit and type of mortar specified using Table 2. The following Articles must be met:

1) Units conform to ASTM C 55 or ASTM C 90 and are sampled and tested in accordance with ASTM C 140.

2) Thickness of bed joints does not exceed $\frac{5}{8}$ in. (16 mm).

3) For grouted masonry, the grout meets one of the following requirements:

a) Grout conforms to ASTM C 476.

b) Grout compressive strength equals f'_m but compressive strength is not less than 2000 psi (13.8 MPa). Determine compressive strength of grout in accordance with ASTM C 1019.

Table 2—Compressive strength of masonry based on the compressive strength of concrete masonry units and type of mortar used in construction

Net area compressive strength of concrete masonry units, psi (MPa)		Net area compressive strength of masonry, psi[1] (MPa)
Type M or S mortar	Type N mortar	
1250 (8.6)	1300 (9.0)	1000 (6.9)
1900 (13.1)	2150 (14.8)	1500 (10.3)
2800 (19.3)	3050 (21.0)	2000 (13.8)
3750 (25.8)	4050 (27.9)	2500 (17.2)
4800 (33.1)	5250 (36.2)	3000 (20.1)

[1] For units of less than 4 in. (102 mm) height, 85 percent of the values listed.

3. *Prism test method*

a. Determine the compressive strength of masonry by the prism test method for the following conditions:

1) When required.

2) When masonry does not meet the requirements for application of the unit strength method.

b. A prism test consists of testing three prisms in accordance with ASTM E 447 Method B modified as follows:

1) Construct prisms in stack bond, one unit long and thick with a full mortar bed.

2) Construct clay masonry prisms with height to thickness ratios in the range from 2.0 to 5.0.

3) Construct concrete masonry prisms with height to thickness ratios in the range from 1.33 to 5.0.

4) Provide a minimum of one joint in hollow unit masonry prisms.

c. The compressive strength of masonry is the average strength of three prisms, but not more than the strength of the masonry units used in prism construction.

d. For clay masonry, multiply the compressive strength of masonry by the height to thickness correction factor indicated in Table 3.

Table 3—Correction factors for clay masonry prism strength

Prism height to thickness ratio	2.0	2.5	3.0	3.5	4.0	4.5	5.0
Correction factor	0.82	0.85	0.88	0.91	0.94	0.97	1.0

e. For concrete masonry, multiply the compressive strength of masonry by the height to thickness correction factor indicated in Table 4.

Table 4—Correction factors for concrete masonry prism strength

Prism height to thickness ratio	1.33	2.0	3.0	4.0	5.0
Correction factor	0.75	1.00	1.07	1.15	1.22

1.5—Submittals

A. *General*

1. Submit the material samples, shop drawings, and documentation as required by this Specification.

2. Obtain written acceptance of submittals prior to the use of the materials or methods requiring acceptance.

B. When required, submit the following:

1. Type and proportions of the ingredients composing the grout mixture(s) to be used in construction.

2. *Samples*

a. Specimens of the masonry units that will be used in project construction, showing range of colors, textures, finishes and dimensions, as well as specimens of colored mortar.

b. One sample, at least 6 in. (152 mm) long, of each type of non-masonry joint material specified.

3. *Test results*

a. Results of mortar tests performed in accordance with the property specification of ASTM C 270.

b. Results of tests of masonry units and materials attesting compliance with the specified requirements.

4. *Construction procedures*

a. Cold weather construction procedures for meeting the requirements of the Project Specifications.

b. Hot weather construction procedures for meeting the requirements of the Project Specifications.

5. Manufacturer's literature.

6. Shop drawings showing:

a. Fabrication dimensions and locations for placement of the reinforcing steel and accessories.

b. Details of steel reinforcement.

c. Lintels and door frames.

d. Shelf angles and lintels.

7. Certification of compliance for:

a. Each type and size of reinforcement to be used in construction, demonstrating compliance with this Specification.

b. Each type and size of anchors, ties, and metal accessories, demonstrating compliance with this Specification and when required, samples of these items.

c. Letter certifying that cement conforms to the requirements of the Contract Documents.

8. Weight slips for grout materials at time of delivery.

1.6—Quality assurance

A. *Testing laboratory services*

1. *General*

a. When required, employ an acceptable independent testing laboratory to perform tests to document submittals, certify product compliance prior to use in construction, establish mortar and grout mix designs, provide supporting data for changes requested by the Contractor, or appeal rejection of material found defective by Owner's tests.

b. The Work will be inspected and evaluated for compliance with the Contract Documents. Unless otherwise required, these services will be paid for by the Owner.

c. Permit and facilitate access of the Owner's representatives to the construction sites and the performance of all activities for quality assurance by these representatives, including inspection and testing required in this Specification.

d. Failure to detect any defective work or material does not in any way prevent later rejection when a defect is discovered and it does not obligate the Architect/Engineer for final acceptance.

2. *Duties and authorities of testing agency designated by the Owner*

a. Representatives of the agency shall inspect, sample, and test the material and shall inspect the construction of masonry in accordance with the Contract Documents. When there is reason to believe that any material furnished or work performed by the Contractor fails to fulfill the requirements of the Contract Documents, report such deficiency to the Architect/Engineer and to the Contractor.

b. The agency shall report all test and inspection results to the Architect/Engineer and Contractor immediately after they are performed. Include in test reports a summary of conditions under which test specimens were stored prior to testing and state what portion of the construction is represented by each test.

c. The testing agency and its representatives are neither authorized to revoke, alter, relax, enlarge, or release any requirement of the Contract Documents, nor authorized to approve or accept, reject or disapprove, any portion of the Work.

3. *Responsibilities and duties of Contractor*

a. The use of testing services does not relieve the Contractor of the responsibility to furnish materials and construction in full compliance with the Contract Documents.

b. Include in the submittals the results of all testing performed to qualify the materials and to establish mix designs.

c. To facilitate testing and inspection, comply with the following:

1) Furnish any necessary labor to assist the designated testing agency in obtaining and handling samples at the project or other sources of material.

2) Advise the designated testing agency sufficiently in advance of operations to allow for completion of quality tests and for the assignment of personnel.

d. Provide and maintain for the sole use of the testing agency adequate facilities for safe storage and proper curing of test specimens on the project site in accordance with the Contract Documents.

4. *Inspection and testing*—The testing agency designated by the Owner shall:

a. Review and when required, test the Contractor's proposed materials for conformance to the Contract Documents.

b. Review and when required, test the Contractor's proposed mix design of mortar and grout for conformance to the Contract Documents.

c. When required, secure production samples of materials at plants or stock piles and test for conformance to the Contract Documents.

d. When required, conduct strength tests of masonry units, prisms, and materials for conformance to the Contract Documents.

e. When required, review the manufacturer's report for cement, reinforcing steel, wall ties and anchors, and/or conduct laboratory tests of the materials as received for conformance with the Contract Documents.

f. When required, perform other testing or inspection services.

B. *Sample panels*

1. When required, construct sample panels of masonry walls using materials and procedures conforming to the Project Specifications.

a. The minimum sample panel size is 4 ft (1.22 m) square.

b. The acceptable standard for the Work is established by the accepted panel.

2. Unless otherwise required, construct sample panels separate from the Work, and retain sample panels at the job site until all Work has been accepted.

1.7—Delivery, storage and handling

A. Do not use damaged masonry units, damaged components of structure, or damaged packaged material.

B. Protect moisture controlled, concrete masonry units and cementitious materials from precipitation or ground water.

C. Do not use masonry materials that are contaminated.

D. Store different aggregates separately.

E. Protect reinforcement, ties, and metal accessories from permanent distortions and store them off the ground.

1.8—Project conditions

A. *Construction loads*—Do not apply construction loads that exceed the safe superimposed load carrying capacity of the masonry and shores, if used.

B. *Masonry protection*—Cover top of unfinished masonry work to protect it from the weather.

C. *Cold weather construction*

1. Implement the following requirements when:

    a. The ambient temperature falls below 40^0 F $(4.5^0$ C), or;

    b. The temperature of masonry units is below 40^0 F $(4.5^0$ C).

2. Do not lay masonry units having a temperature below 20^0 F $(-7^0$ C). Remove visible ice on masonry units before the unit is laid in the masonry.

3. Heat mortar sand or mixing water to produce mortar temperatures between 40^0 F $(4.5^0$ C) and 120^0 F $(49^0$ C) at the time of mixing. Maintain mortar above freezing until used in masonry.

4. When ambient temperature is between 25^0 F $(-4^0$ C) and 20^0 F $(-7^0$ C) use heat sources on both sides of the masonry under construction and install wind breaks when wind velocity is in excess of 15 mph (24 km/h).

5. When ambient temperature is below 20^0 F $(-7^0$ C), provide an enclosure for the masonry under construction and use heat sources to maintain temperatures above 32^0 F $(0^0$ C) within the enclosure.

6. When mean daily temperature is between 40^0 F $(4.5^0$ C) and 32^0 F $(0^0$ C), protect completed masonry from rain or snow by covering with a weather resistive membrane for 24 hr after construction.

7. When mean daily temperature is between 32^0 F $(0^0$ C) and 25^0 F $(-4^0$ C), completely cover completed masonry with a weather resistive membrane for 24 hr after construction.

8. When mean daily temperature is between 25^0 F $(-4^0$ C) and 20^0 F $(-7^0$ C), completely cover completed masonry with insulating blankets or equal protection for 24 hr after construction.

9. When mean daily temperature is below 20^0 F $(-7^0$ C), maintain masonry temperature above 32^0 F $(0^0$ C) for 24 hr after construction by enclosure with supplementary heat, by electric heating blankets, by infrared heat lamps or by other acceptable methods.

10. Do not lay glass unit masonry during cold weather construction periods as defined in Article 1.8C1 a or b. Maintain temperature of glass unit masonry above 40^0 F $(4.5^0$ C) for the first 48 hr after construction.

D. *Hot weather construction*

1. Implement the following when the ambient air temperature exceeds the following:

    a. 100^0 F $(38^0$ C), or;

    b. 90^0 F $(32^0$ C) with a wind velocity greater than 8 mph (13 km/h).

2. Do not spread mortar beds more than 4 ft (1.2 m) ahead of masonry.

3. Set masonry units within one min. of spreading mortar.

PART 2—PRODUCTS

2.1—Mortar materials

A. Provide mortar of the type and color specified that conforms to ASTM C 270.

B. *Glass unit masonry*

1. Provide Type S or N mortar that conforms to Article 2.1A.

2. Comply with the other requirements of Article 2.6.

2.2—Grout materials

A. Unless otherwise required, provide grout that conforms to the requirements of ASTM C 476. Do not use admixtures unless acceptable.

2.3—Masonry materials

A. Provide concrete masonry units that conform to ASTM C 55, C 73, C 90, C 129, or C 744 as specified.

B. Provide clay or shale masonry units that conform to ASTM C 34, C 56, C 62, C 126, C 212, C 216, or C 652, as specified.

C. Provide stone masonry units that conform to ASTM C 503, C 568, C 615, C 616 or C 629, as specified.

D. Provide hollow glass units that are partially evacuated and have a minimum average glass face thickness of $3/16$ in. (4.8 mm). Provide solid glass block units when required. Provide units in which the surfaces intended to be in contact with mortar are treated with polyvinyl butyral coating or latex-based paint. Do not use reclaimed units.

2.4—Reinforcement and metal accessories

A. *Reinforcing steel*—Provide deformed reinforcing bars that conform to one of the following as specified:

1. ASTM A 615.
2. ASTM A 616, including Supplement 1.
3. ASTM A 617.
4. ASTM A 706.
5. ASTM A 767.
6. ASTM A 775.

B. *Joint reinforcement*

1. Except as otherwise specified, provide joint reinforcement manufactured with wire conforming to ASTM A 82 and with deformed longitudinal wires. One set of deformations shall occur around the perimeter of the wire at a maximum spacing of 0.7 times the diameter of the wire but there shall not be less than eight sets per inch of length. The overall length of each deformation within the set shall be such that the sum of gaps between the ends of the deformations shall not exceed 25 percent of the perimeter of the wire. The height or depth of the deformations shall be 0.012 in. (0.30 mm) for wire size W2.8 or larger, 0.011 in. (0.28 mm) for wire size W2.1, and 0.009 in. (0.23 mm) for wire size W1.7. Cross wires are permitted to be plain.

2. *Deformed reinforcing wire*—Provide deformed reinforcing wire that conforms to ASTM A 496.

3. *Welded wire fabric*—Provide welded wire fabric that conforms to one of the following specifications:

    a. Plain . . . . ASTM A 185
    b. Deformed . . . . ASTM A 497

C. *Anchors, ties and accessories*—Provide anchors, ties, and accessories that conform with the following specifications, except as otherwise specified:

1. Plate and bent bar anchors . . . . . ASTM A 36
2. Sheet metal anchors and ties . . ASTM A 366
3. Wire mesh ties . . . . ASTM A 185
4. Wire ties and anchors . . . . ASTM A 82
5. Anchor bolts . . . . . . ASTM A 307, Grade A
6. Panel anchors (for glass unit masonry)—Provide $1\,3/4$ in. (19 mm) wide, 24 in. (610 mm) long, 20 gage steel strips, punched with three staggered rows of elongated holes, galvanized after fabrication.

D. *Stainless steel*—When required, provide type 304 stainless steel items that comply with the following specifications:

1. Joint reinforcement . . . . ASTM A 580
2. Plate and bent bar anchors . . . . ASTM A 666
3. Sheet metal anchors and ties . . ASTM A 167
4. Wire ties and anchors . . . . ASTM A 580

E. *Coatings for corrosion protection*—Unless otherwise required, protect carbon steel joint reinforcement, ties, and anchors from corrosion by galvanizing in conformance with the following minimum.

1. Joint reinforcement, interior walls
. . . . ASTM A 641(0.1 oz/ft^2) (0.031 kg/m^2)

2. Joint reinforcement, wire ties, or wire anchors in exterior walls or interior walls exposed to a mean relative humidity exceeding 75 percent (e.g. natatoria and food processing)
. . . . ASTM A 153 (1.50 oz/ft^2) (0.46 kg/m^2)

3. Sheet metal ties or anchors in exterior walls or interior walls exposed to a mean relative humidity exceeding 75 percent (e.g. natatoria and food processing)
. . . . ASTM A 153 Class B

4. Sheet metal ties or anchors in interior walls
. . . . ASTM A 525 Class G60

5. Steel plates and bars (as applicable to size and form indicated) . . . . ASTM A 123 or
. . . . ASTM A 153, Class B

2.5—Accessories

A. Unless otherwise required, provide contraction joint material that conforms to one of the following standards:

1. ASTM D 2000, M2AA-805 Rubber shear keys with a minimum durometer hardness of 80.
2. ASTM D 2287, Type PVC 654-4 PVC shear keys with a minimum durometer hardness of 85.
3. ASTM C 920.

B. Unless otherwise required, provide expansion joint material that conforms to one of the following standards:

1. ASTM C 920.
2. ASTM D 994.
3. ASTM D 1056, Class 2A1.

C. *Masonry cleaner*

1. Use potable water and detergents to clean masonry unless otherwise acceptable.

2. Unless otherwise required, do not use acid or caustic solutions.

D. *Joint fillers*—Use the size and shape of joint fillers specified.

2.6—Mixing

A. *Mortar*

1. Mix all cementitious materials and aggregates between 3 and 5 min. in a mechanical batch mixer with a sufficient amount of water to produce a workable consistency. Unless acceptable, do not hand mix mortar. Maintain workability of mortar by remixing or retempering. Discard all mortar which has begun to stiffen or is not used within $2^1/_2$ hours after initial mixing.

2. When required, use mineral oxide or carbon black job site pigments. Limit the maximum percentage by weight of cement as follows:

    a. Pigmented portland cement-lime mortar
        1) Mineral oxide pigment . . . . 10 percent
        2) Carbon black pigment . . . . . 2 percent
    b. Pigmented masonry cement mortar
        1) Mineral oxide pigment . . . . . 5 percent
        2) Carbon black pigment . . . . . 1 percent

3. Do not use admixtures containing more than 0.2 percent chloride ions.

4. *Glass unit masonry*—Reduce the amount of water to account for the lack of absorption. Do not retemper mortar after initial set. Discard unused mortar within $1^1/_2$ hr after initial mixing.

B. *Grout*

1. Unless otherwise required, proportion and mix grout in accordance with the requirements of ASTM C 476.

2. Unless otherwise required, mix grout to a consistency that has a slump between 8 and 11 in. (203 and 279 mm). When tests are required, test grout slump in accordance with ASTM C 143.

2.7—Fabrication

A. *Reinforcement*

1. Fabricate bars used in masonry reinforcement in accordance with the fabricating tolerances of ACI 315.

2. Unless otherwise required, bend bars cold and do not heat bars.

3. The minimum inside diameter of bend for stirrups shall be five bar diameters.

4. Do not bend grade 40 bars in excess of 180 deg. The minimum inside diameter of bend is five bar diameters.

5. The minimum inside bend diameter for other bars is as follows:

    a. No. 3 through No. 8 . . . . . 6 bar diameters
    b. No. 9 through No. 11 . . . . 8 bar diameters

6. Provide standard hooks that conform to the following:

    a. A standard 180 deg hook: 180 deg bend plus a minimum extension of 4 bar diameters or $2^1/_2$ in. (64 mm), whichever is greater.

    b. A standard 135 deg hook: a 135 deg bend plus a minimum extension of 6 bar diameters or 4 in. (102 mm), whichever is greater.

    c. A standard 90 deg hook: 90 deg bend plus a minimum extension of 12 bar diameters.

    d. For stirrups: a 90 or 135 deg bend plus a minimum of 6 bar diameters or $2^1/_2$ in. (64 mm), whichever is greater.

7. Fabricate joint reinforcement, anchors, and ties in accordance with the Contract Documents and with the published specifications of the accepted manufacturer.

B. *Prefabricated masonry*

1. Unless otherwise required, provide prefabricated masonry that conforms to the provisions of ASTM C 901 and the Contract Documents.

2. When used, provide prefabricated masonry that conforms to the details shown on the Project Drawings and the requirements in the Contract Documents.

3. Unless otherwise required, provide prefabricated masonry lintels that have an appearance similar to the masonry units used in the wall surrounding each lintel.

4. Mark prefabricated masonry for proper location and orientation.

2.8—Source quality

A. When tests are required, perform tests of masonry units in accordance with the following standards:

    1. Concrete masonry . . . . ASTM C 140
    2. Clay masonry . . . . ASTM C 67
    3. Stone masonry
        a. Structural granite, building sandstone, marble, building stone (exterior), limestone building stone
        . . . . ASTM C 97, C 99, and C 170
        b. Structural slate . . ASTM C 120, and C 121

PART 3—EXECUTION

3.1—Inspection

A. Prior to the start of masonry construction, the Contractor shall verify:

1. That foundations are constructed with tolerances conforming to the requirements of ACI 117.

2. That reinforcing dowels are positioned in accordance with the Project Drawings.

3. If stated conditions are not met, notify the Architect/Engineer.

3.2—Preparation

A. Clean all reinforcement by removing mud, oil, or other materials that will adversely affect or reduce the bond at the time mortar or grout is placed. Reinforcement with rust, mill scale, or a combination of both will be accepted as being satisfactory without cleaning or brushing provided the dimensions and weights, including heights of deformations, of a cleaned sample are not less than required by the ASTM specification covering this reinforcement in this Specification.

B. Prior to placing masonry, remove laitance, loose aggregate, and anything else that would prevent mortar from bonding to the foundation.

C. *Wetting masonry units*

1. *Concrete masonry*—Unless otherwise required, do not wet concrete masonry units before laying.

2. *Clay or shale masonry*—Wet clay or shale masonry units having initial absorption rates in excess of one gram per min. per in.^{2} (0.0016 grams per min. per mm^2), when measured in accordance with ASTM C 67, so the initial rate of absorption will not exceed one gram per min. per in.^{2} (0.0016 grams per min, per mm^2) when the units are used. Lay wetted units when surface dry. Do not wet clay or shale masonry units having an initial absorption 0.00031 rate less than 0.2 grams per min. per in.^{2} (0.00031 grams per min. per mm^2).

D. *Debris*—Construct grout spaces free of mortar dropping, debris, loose aggregates, and any material deleterious to masonry grout.

E. *Reinforcement*—Place reinforcement and ties in grout spaces prior to grouting.

F. *Cleanouts*—Provide cleanouts in the bottom course of masonry for each grout pour, when the grout pour height exceeds 5 ft (1.5 m).

1. Where required, construct cleanouts adjacent to each vertical bar. In solid grouted masonry, space cleanouts horizontally a maximum of 32 in. (813 mm) on center.

2. Construct cleanouts with an opening of sufficient size to permit removal of debris. The minimum opening dimension shall be 3 in. (76 mm).

3. After cleaning, close cleanouts with closures braced to resist grout pressure.

3.3—Masonry erection

A. *Bond pattern*—Unless otherwise required, construct masonry in running bond pattern.

B. *Placing mortar and units*

1. *Bed and head joints*—Unless otherwise required or indicated on the Project Drawings, construct $^3/_8$ in. (9.5 mm) thick bed and head joints except at foundation or with glass unit masonry. Construct bed joint of the starting course of foundation with a thickness not less than $^1/_4$ in. (6.4 mm) and not more than $^3/_4$ in. (19 mm). Provide glass unit masonry bed and head joint thicknesses in accordance with Article 3.3B5b. Construct joints that also conform to the following:

a. Fill holes not specified in exposed and below grade masonry with mortar.

b. Unless otherwise required, tool joint with a round jointer when the mortar is thumbprint hard.

c. Remove masonry protrusions extending $^1/_2$ in. (13 mm) or more into cells or cavities to be grouted.

2. *Collar joints*—Unless otherwise required or indicated on the Project Drawings, solidly fill collar joints less than $^3/_4$ in. (19 mm) wide with mortar as the job progresses.

3. *Hollow units*—Place hollow units so:

a. Face shells of bed joints are fully mortared.

b. Webs are fully mortared in all courses of piers, columns and pilasters, in the starting course on foundations, when necessary to confine grout or loose-fill insulation, and when otherwise required.

c. Head joints are mortared, a minimum distance from each face equal to the face shell thickness of the unit.

d. Vertical cells to be grouted are aligned and unobstructed openings for grout are provided in accordance with the Project Drawings.

4. *Solid units*—Unless otherwise required, solidly fill bed and head joints with mortar and:

a. Do not fill head joints by slushing with mortar.

b. Construct head joints by shoving mortar tight against the adjoining unit.

c. Do not deeply furrow bed joints.

5. *Glass units*

a. Lay units so head and bed joints are filled solidly. Do not furrow mortar.

b. Unless otherwise required, construct head and bed joints of glass unit masonry $^1/_4$ in. (64 mm) thick, except that vertical joint thickness of radial panels shall not be less than $^1/_8$ in. (3.2 mm). The bed joint thickness tolerance shall be minus $^1/_{16}$ in. (1.6 mm) and plus $^1/_8$ in. (3.2 mm). The head joint thickness tolerance shall be plus or minus $^1/_8$ in. (3.2 mm).

c. Comply with the Project Specifications for other installation requirements.

6. *All units*

a. Place clean units while the mortar is soft and plastic. Remove and relay in fresh mortar any unit disturbed to the extent that initial bond is broken after initial positioning.

b. Cut exposed edges or faces of masonry units smooth or position such that all exposed faces or edges are unaltered manufactured surfaces.

    c. When the bearing of a masonry wythe on its support is less than two-thirds of the wythe thickness, notify the Architect/Engineer.

C. *Prefabricated concrete and masonry items*—Erect prefabricated concrete and masonry items in accordance with the requirements indicated on the Project Drawings.

D. *Embedded items and accessories*—Install embedded items and accessories where shown in the Project Drawings and in accordance with the Contract Documents.

    1. Construct chases as masonry units are laid.

    2. When required, place pipes and conduits passing horizontally through masonry beams or masonry walls in steel sleeves or cored holes.

    3. Install pipes and conduits passing horizontally through nonbearing masonry partitions.

    4. Place pipes and conduits passing horizontally through piers, pilasters, or columns.

    5. Place horizontal pipes and conduits in and parallel to plane of walls.

    6. Install and secure connectors, flashing, weep holes, weep vents, nailing blocks, and other accessories.

    7. Install movement joints.

    8. When required, install insulation.

    9. Aluminum—Do not embed aluminum conduits, pipes, and accessories in masonry, grout, or mortar, unless effectively coated or covered to prevent aluminum cement chemical reaction or electrolytic action between aluminum and steel.

E. *Bracing of masonry*—Design, provide, and install bracing that will assure stability of masonry during construction.

F. *Site tolerances*—Erect masonry within the following tolerances from the specified dimensions.

    1. Dimension of elements

      a. In cross section or elevation

        . . . . $-\frac{1}{4}$ in. (6.4 mm), $+\frac{1}{2}$ in. (13 mm)

      b. Mortar joint thickness

        bed . . . . $\pm\frac{1}{8}$ in. (3.2 mm)

        head . . . $-\frac{1}{4}$ in. (6.4 mm), $+\frac{3}{8}$ in. (9.5 mm)

        collar . $-\frac{1}{4}$ in. (6.4 mm), $+\frac{3}{8}$ in. (9.5 mm)

        glass unit masonry . . . . see Article 3.3B5b

      c. Grout space or cavity width, except for masonry walls passing framed construction

        . . . . . . $-\frac{1}{4}$ in. (6.4 mm), $+\frac{3}{8}$ in. (9.5 mm)

    2. Elements

      a. Variation from level:

        bed joints

        . . . . . . $\pm\frac{1}{4}$ in. (6.4 mm) in 10 ft (3.1 m)

        . . . . $\pm\frac{1}{2}$ in. (13 mm) maximum

        top surface of bearing walls

        . . . . . . $\pm\frac{1}{4}$ in. (6.4 mm) in 10 ft (3.1 m)

        . . . . $\pm\frac{1}{2}$ in. (13 mm) maximum

      b. Variation from plumb

        . . . . . . $\pm\frac{1}{4}$ in. (6.4 mm) in 10 ft (3.1 m)

        . . . . . . $\pm\frac{3}{8}$ in. (9.5 mm) in 20 ft (6.1 m)

        . . . . $\pm\frac{1}{2}$ in. (13 mm) maximum

      c. True to a line

        . . . . . . $\pm\frac{1}{4}$ in. (6.4 mm) in 10 ft (3.1 m)

        . . . . . . $\pm\frac{3}{8}$ in. (9.5 mm) in 20 ft (6.1 m)

        . . . . $\pm\frac{1}{2}$ in. (13 mm) maximum

      d. Alignment of columns and walls (bottom versus top)

        . . . . . . $\pm\frac{1}{2}$ in. (13 mm) for bearing walls

        . . . $\pm\frac{3}{4}$ in. (19 mm) for nonbearing walls

    3. Location of elements

      a. Indicated in plan

        . . . . . . $\pm\frac{1}{2}$ in. (13 mm) in 20 ft (6.1 m)

        . . . . $\pm\frac{3}{4}$ in. (19 mm) maximum

      b. Indicated in elevation

        . . . . . . $\pm\frac{1}{4}$ in. (6.4 mm) in story height

        . . . . $\pm\frac{3}{4}$ in. (19 mm) maximum

    4. If the above conditions cannot be met due to previous construction, notify the Architect/ Engineer.

3.4—Reinforcement installation

A. *Basic requirements*—Place reinforcement and accessories in accordance with the sizes, types, and locations indicated on the Project Drawings, and as specified. Do not place dissimilar metals in contact with each other.

B. *Securing reinforcement*—Support and fasten reinforcement together to prevent displacement by construction loads or by placement of grout or mortar beyond the tolerances allowed.

C. *Details of reinforcement*

    1. Maintain clear distance between reinforcing bars and any face of masonry unit or formed surface, as indicated on the Project Drawings, but not less than $\frac{1}{4}$ in. (6.4 mm) for fine grout or $\frac{1}{2}$ in. (13 mm) for coarse grout.

    2. Splice only where indicated on the Project Drawings, unless otherwise acceptable.

    3. Unless accepted by the Architect/Engineer, do not bend reinforcement after it is embedded in grout or mortar.

    4. Place joint reinforcement so that longitudinal wires are embedded in mortar with a minimum cover of $\frac{1}{2}$ in. (13 mm) when not exposed to weather or earth and $\frac{5}{8}$ in. (16 mm) when exposed to weather or earth.

D. *Wall ties*

    1. Embed the ends of wall ties in mortar joints. Embed wall tie ends at least $\frac{1}{2}$ in. (13 mm) into the outer face shell of hollow units. Embed wire wall ties at least $1\frac{1}{2}$ in. (38 mm) into the mortar bed of solid masonry units or solid grouted hollow units.

    2. Unless otherwise required, bond wythes not bonded by headers with wall ties as follows:

Wire size	Minimum number of ties required
W1.7	One wall tie wire per 2.67 ft² (0.25 m²)
W2.8	One wall tie wire per 4.50 ft² (0.42 m²)

The maximum spacing between ties is 36 in. (914 mm) horizontally and 24 in. (610 mm) vertically.

    3. Unless accepted by the Architect/Engineer, do not bend wall ties after being embedded in grout or mortar.

4. Unless otherwise required, install adjustable ties in accordance with the following requirements:

a. One tie for each 1.77 ft² (0.16 m²) of wall area.

b. Do not exceed 16 in. (406 mm) horizontal or vertical spacing.

c. The maximum misalignment of bed joints from one wythe to the other is $1\frac{1}{4}$ in. (32 mm).

d. The maximum clearance between connecting parts of the ties is $\frac{1}{16}$ in. (1.6 mm).

e. When pintle legs are used, provide ties with at least two legs made of wire size W2.8.

5. Install wire ties perpendicular to a vertical line on the face of the wythe from which they protrude. Where one-piece ties or joint reinforcement are used, the bed joints of adjacent wythes shall align.

6. Unless otherwise required, provide additional unit ties around all openings larger than 16 inches (406 mm) in either dimension. Space ties around perimeter of opening at a maximum of 3 ft (0.9 m) on center. Place ties within 12 inches (305 mm) of opening.

E. *Site tolerances*

1. Tolerances for the placement of steel in walls and flexural elements shall be $\pm \frac{1}{2}$ in. (13 mm) when the distance from the centerline of steel to the opposite face of masonry, d, is equal to 8 in. (203 mm) or less, ± 1 in. (25 mm) for d equal to 24 in. (610 mm) or less but greater than 8 in. (203 mm), and $\pm 1\frac{1}{4}$ in. (32 mm) for d greater than 24 in. (610 mm).

2. In walls, for vertical bars, 2 in. (51 mm) from the location along the length of the wall indicated on the Project Drawings.

3. If it is necessary to move bars more than one bar diameter or a distance exceeding the tolerance stated above to avoid interference with other reinforcing steel, conduits, or embedded items, notify the Architect/Engineer for acceptance of the resulting arrangement of bars.

F. *Glass unit masonry panel anchors*—When used in lieu of channel-type restraints, install panel anchors as follows:

1. Unless otherwise required, space panel anchors at 16 in. (406 mm) in both the jambs and across the head.

2. Embed panel anchors a minimum of 12 in. (305 mm).

3. Provide two fasteners, capable of resisting the required loads, per panel anchor.

3.5—Grout placement

A. *Placing time*—Place grout within $1\frac{1}{2}$ hr from introducing water in the mixture and prior to initial set.

B. *Confinement*—Confine grout to the areas indicated on the Project Drawings. Use material to confine grout that permits bond between masonry units and mortar.

C. *Grout pour height*—Do not exceed the maximum grout pour height given in Table 5.

D. *Grout lift height*—Place grout in lifts not exceeding 5 ft (1.5 m).

E. *Consolidation*—Consolidate grout at the time of placement.

F. Consolidate grout pours 12 in. (305 mm) or less in height by mechanical vibration or by puddling.

G. Consolidate pours exceeding 12 in. (305 mm) in height by mechanical vibration and reconsolidate by mechanical vibration after initial water loss and settlement has occurred.

Table 5—Grout space requirements

Specified grout type[1]	Maximum grout pour height, ft (m)	Minimum width of grout space, in. (mm)[2,3]	Minimum grout space dimensions for grouting cells of hollow units,[3,4] in. x in. (mm x mm)
Fine	1 (0.305)	$\frac{3}{4}$ (19)	$1\frac{1}{2}$ x 2 (38 x 51)
Fine	5 (1.52)	2 (51)	2 x 3 (51 x 76)
Fine	12 (3.66)	$2\frac{1}{2}$ (64)	$2\frac{1}{2}$ x 3 (64 x 76)
Fine	24 (7.32)	3 (76)	3 x 3 (76 x 76)
Coarse	1 (0.305)	$1\frac{1}{2}$ (38)	$1\frac{1}{2}$ x 3 (38 x 76)
Coarse	5 (1.52)	2 (51)	$2\frac{1}{2}$ x 3 (64 x 76)
Coarse	12 (3.66)	$2\frac{1}{2}$ (64)	3 x 3 (76 x 76)
Coarse	24 (7.32)	3 (76)	3 x 4 (76 x 102)

[1] Fine and coarse grouts are defined in ASTM C 476. Grout shall attain a minimum compressive strength of 2000 psi (13.8 MPa) at 28 days.

[2] For grouting between masonry wythes.

[3] Grout space dimension is the clear dimension between any masonry protrusion and shall be increased by the diameters of the horizontal bars within the cross section of the grout space.

[4] Area of vertical reinforcement shall not exceed 6 percent of the area of the grout space.

3.6—Field quality control

A. When the testing of prisms is required, perform one test prior to construction and perform at least one test during construction for each 5000 ft² (465 m²) of wall area or portion thereof.

B. When required, test mortar in accordance with the property specifications of ASTM C 270 or evaluate in accordance with ASTM C 780.

C. When required, the designated testing agency will sample and test grout in accordance with ASTM C 1019 for each 5000 ft² (465 m²) of masonry wall surface.

3.7—Cleaning

A. Clean exposed masonry surfaces of all stains, efflorescence, mortar or grout droppings, and debris.

Commentary on Building Code Requirements for Masonry Structures (ACI 530-95/ASCE 5-95/TMS 402-95)

Reported by the Masonry Standards Joint Committee

James Colville
Chairman

Max L. Porter
Vice Chairman

J. Gregg Borchelt
Secretary

Maribeth S. Bradfield
Membership Secretary

Regular Members[1]:

Gene C. Abbate	Gerald A. Dalrymple	Rochelle C. Jaffe	Joseph F. Neussendorfer
Bechara E. Abboud	Steve Dill	John C. Kariotis	Joseph E. Saliba
Bijan Ahmadi	Russell T. Flynn	Richard E. Klingner	Arturo Schultz
Amde M. Amde	John A. Frauenhoffer	Walter Laska	Matthew J. Scolforo
Richard H. Atkinson	Thomas A. Gangel	L. Donald Leinweber	Daniel Shapiro
William G. Bailey	Richard M. Gensert	Hugh C. MacDonald, Jr.	John M. Sheehan
Stuart R. Beavers	Satyendra K. Ghosh	Billy R. Manning	Robert A. Speed
Robert J. Beiner	Clayford T. Grimm	John H. Matthys	Ervell A. Staab
Frank Berg	John C. Grogan	Robert McCluer	Jerry G. Stockbridge
Russell H. Brown	Craig K. Haney	Donald G. McMican	Itzhak Tepper
A. Dwayne Bryant	Gary C. Hart	George A. Miller	Robert C. Thacker
Kevin D. Callahan	Barbara Heller	Reg Miller	Donald W. Vannoy
Mario J. Catani	Robert Hendershot	Colin C. Munro	Terence A. Weigel
Robert W. Crooks	Mark B. Hogan	W. Thomas Munsell	A. Rhett Whitlock
Kenneth G. Dagostino, Jr.	Thomas A. Holm	Antonio Nanni	

Associate Members[2]:

James E. Amrhein	Hans R. Ganz	Robert F. Mast	Phillip J. Samblanet
David T. Biggs	H. R. Hamilton, III	John Melander	Richard C. Schumacher
James W. Cowie	B. A. Haseltine	Raul Alamo Neihart	John G. Tawresey
John Chrysler	Edwin G. Hedstrom	Robert L. Nelson	Robert D. Thomas
Terry M. Curtis	A. W. Hendry	Rick Okawa	Dean J. Tills
Walter L. Dickey	Thomas F. Herrell	Adrian W. Page	Charles W. C. Yancey
Jeffrey L. Elder	Steve Lawrence	Ruiz Lopez M. Rafael	
Brent A. Gabby	Nicholas T. Loomis	Roscoe Reeves, Jr.	

SYNOPSIS

This commentary documents some of the considerations of the Masonry Standards Joint Committee in developing the provisions contained in "Building Code Requirements for Masonry Structures (ACI 530-95/ ASCE 5-95/TMS 402-95)." This information is provided in the commentary because this Code is written as a legal document and cannot therefore present background details or suggestions for carrying out its requirements.

Emphasis is given to the explanation of new or revised provisions that may be unfamiliar to users of this Code. References to much of the research data used to prepare this Code are cited for the user desiring to study individual items in greater detail. The subjects covered are those found in this Code. The chapter and section numbering of this Code are followed throughout.

Keywords: anchors (fasteners); anchorage (structural); beams; **building codes**; cements; clay brick; clay tile; columns; compressive strength; concrete block; concrete brick; **construction**; detailing; empirical design flexural strength; glass units; grout; grouting; joints; loads (forces); **masonry**; masonry cements; masonry load-bearing walls; masonry mortars; masonry walls; modulus of elasticity; mortars; pilasters; quality assurance; reinforced masonry; reinforcing steel; seismic requirements; shear strength; specifications; splicing; stresses; structural analysis; structural design; ties; unreinforced masonry; veneers (anchored); walls; working stress design.

---

[1]Regular members fully participate in Committee activities, including responding to correspondence and voting.

[2]Associate members monitor Committee activities, but do not have voting privileges.

CONTENTS

INTRODUCTION

This commentary documents some of the considerations of the Masonry Standards Joint Committee in developing the provisions contained in "Building Code Requirements for Masonry Structures (ACI 530-95/ASCE 5-95/TMS 402- 95)," hereinafter called this Code. Comments on specific provisions are made under the corresponding chapter and section numbers of this Code.

The commentary is not intended to provide a detailed account of the studies and research data reviewed by the committee in formulating the provisions of this Code. However, references to some of the research data are provided for those who wish to study the background material in depth.

As the name implies, "Building Code Requirements for Masonry Structures (ACI 530-95/ASCE 5-95/TMS 402- 95)" is meant to be used as part of a legally adopted building code and as such must differ in form and substance from documents that provide detailed specifications, recommended practices, complete design procedures, or design aids.

This Code is intended to cover all buildings of the usual types, both large and small. This Code and this commentary cannot replace sound engineering knowledge, experience, and judgment. Requirements more stringent than this Code provisions may sometimes be desirable.

A building code states only the minimum requirements necessary to provide for public health and safety. The ACI-ASCE-TMS Building Code is based on this principle. For any structure, the owner or the structural designer may require the quality of materials and construction to be higher than the minimum requirements necessary to protect the public as stated in this Code. However, lower standards are not permitted.

This commentary directs attention to other documents that provide suggestions for carrying out the requirements and intent of this Code. However, those documents and this commentary are not intended to be a part of this Code.

This Code has no legal status unless it is adopted by government bodies having the police power to regulate building design and construction or unless incorporated into a contract. Where this Code has not been adopted, it may serve as a reference to good practice even though it has no legal status.

This Code provides a means of establishing minimum standards for acceptance of designs and construction by a legally appointed Building Official or his designated representatives. Therefore, this Code cannot define the contract responsibility of each of the parties in usual construction unless incorporated into a contract. However, general references requiring compliance with this Code in the project specifications are improper since minimum code requirements should be incorporated in the contract documents which should contain all requirements necessary for construction.

Masonry is one of the oldest forms of construction. In modern times the design of masonry has been governed by standards which separate clay masonry from concrete masonry. For this Code, the committee has adopted the policy that the design methodology for all masonry should be the same. The committee adopted this policy in recognition that the design methodology developed does not always predict the actual performance of masonry as accurately as it would like and that masonry work designed in accordance with some empirical provisions performs better than would be indicated by current design procedures. These design situations are being identified by the committee and singled out for further detailed research.

PART 1—GENERAL

CHAPTER 1—GENERAL REQUIREMENTS

1.1—Scope

This Code covers the structural design and construction of masonry elements and serves as a part of the general building code. Since the requirements for masonry in this Code are interrelated, this Code may need to supersede when there are conflicts on masonry design and construction with the general building code or with documents referenced by this Code. The designer must resolve the conflict for each specific case.

1.2—Contract Documents and calculations

1.2.1 The provisions for preparation of project drawings, project specifications, and issuance of permits are, in general, consistent with those of most general building codes and are intended as supplements thereto.

1.2.2 This Code lists some of the more important items of information that must be included in the project drawings or project specifications. This is not an all inclusive list and additional items may be required by the Building Official.

1.2.3 The contract documents must accurately reflect design requirements. For example, joint and opening locations assumed in the design should be coordinated with locations shown on the drawings.

Verifications that masonry construction conforms to the contract documents is required by this Code. A program of quality assurance must be included in the contract documents to satisfy this Code requirement.

1.2.4 This Code accepts documented computer programs as a means of obtaining a structural analysis or design in lieu of detailed manual calculations. The extent of input and output information required will vary according to the specific requirements of individual Building Officials. However, when a computer program has been used by the designer, only skeleton data should normally be required. Design assumptions and program documentation are necessary. This should consist of sufficient input and output data and other information to allow the Building Official to perform a detailed review and make comparisons using another program or manual calculations. Input data should be identified as to member designation, applied loads, and span lengths. The related output data should include member designation and the shears, moments, and reactions at key points. Recommendations for computer submittals are detailed in "Recommended Documentation for Computer Calculation Submittals to Building Officials" reported by ACI Committee 118.

1.3—Approval of special systems of design and construction

New methods of design, new materials, and new uses of materials must undergo a period of development before being specifically covered in a code. Hence, valid systems or components might be excluded from use by implication if means were not available to obtain acceptance. This section permits proponents to submit data substantiating the adequacy of their system or component to a "board of examiners." Such a board should be created and named in accordance with local laws, and should be headed by a registered engineer. All board members should be directly associated with, and competent in, the fields of structural design or construction of masonry.

For special systems considered under this section, specific tests, load factors, deflection limits, and other pertinent requirements should be set by the board of examiners, and should be consistent with the intent of the code.

1.4—Standards cited in this Code

These standards are referenced in this Code. Specific dates are listed here since changes to the standard may result in changes of properties or procedures.

CHAPTER 2—NOTATIONS AND DEFINITIONS

2.1—Notations

Notations used in this Code are summarized here. Each symbol is unique, with the notation as used in other masonry standards when possible.

2.2—Definitions

For consistent application of this Code, terms are defined which have particular meanings in this Code. The definitions given are for use in application of this Code only and do not always correspond to ordinary usage. Glossaries of masonry terminology are available from several sources within the industry.[2.1,2.2,2.3]

The permitted tolerance for units are found in the appropriate materials standards. Permitted tolerances for joints and masonry construction are found in the Specification. Nominal dimensions are usually used to identify the size of a masonry unit. The thickness or width is given first, followed by height and length. Nominal dimensions are normally given in whole numbers nearest to the specified dimensions. Specified dimensions are most often used for design calculations.

References

2.1. "Glossary of Terms Relating to Brick Masonry," *Technical Notes on Brick Construction*, No. 2 (Revised), Brick Institute of America, Reston, VA, 1988, 4 pp.

2.2. "Glossary of Concrete Masonry Terms," *NCMA TEK Bulletin* No. 145, National Concrete Masonry Association, Herndon, VA, 1985, 4 pp.

2.3. "The Masonry Glossary," International Masonry Institute, Washington, D.C., 1981, 144 pp.

PART 2—QUALITY ASSURANCE AND CONSTRUCTION REQUIREMENTS

CHAPTER 3—GENERAL

3.1—Materials, labor, and construction

3.1.1 The ACI 530.1/ASCE 6/TMS 602 Specification covers material and construction requirements. It is an integral part of this Code in terms of minimum requirements relative to the composition, quality, storage, handling, and placement of materials for masonry structures. The Specification also includes provisions requiring verification that construction achieves the quality specified. The construction must conform to these requirements. A quality assurance program must be defined in the contract documents. Since the design and the complexity of the masonry construction varies from project to project, so must the extent of the quality assurance program. The quality assurance program must include quality control requirements. Depending on the scope of the masonry work, the quality assurance program may include field inspection and testing requirements. The contract documents must indicate the testing, inspection, and other measures that are required to ensure that the Work is in conformance with the project requirements.

An Owner's representative must be identified. The Owner's representative will evaluate test results and perform the inspections during construction. The Owner's representative may be a third party, or the design professional specifically employed to perform those duties.

This Code is not intended to be made a part of the contract documents. The contractor should not be asked through contract documents to assume responsibility regarding design (Code) requirements. A commentary on ACI 530.1/ASCE 6/TMS 602 follows this commentary.

3.1.2 Code Table 3.1.2 contains the least clear dimension for grouting between wythes and the minimum cell dimensions when grouting hollow units. Selection of units and bonding pattern should be coordinated to achieve these requirements. Vertical alignment of cells must also be considered. All projections or obstructions into the grout space and the diameter of horizontal reinforcement must be considered when calculating the minimum dimensions. See Fig. 3.1-1.

Coarse grout and fine grout are differentiated by aggregate size in ASTM C 476.

The grout space requirements of Code Table 3.1.2 are based on usual grout aggregate size and cleaning practice to permit complete filling of grout spaces and adequate consolidation using typical methods of construction. Grout spaces smaller than specified have been used successfully in some areas, however, the committee felt that these more conservative requirements should be required. Application for a variance to Code provisions can be made where construction practice is such that smaller spaces can be grouted successfully.

3.2—Acceptance relative to strength requirements

Fundamental to the structural adequacy of masonry construction is the necessity that the compressive strength of masonry equals or exceeds the specified strength. Rather than mandating design based on different values of f'_m for each wythe of a multiwythe

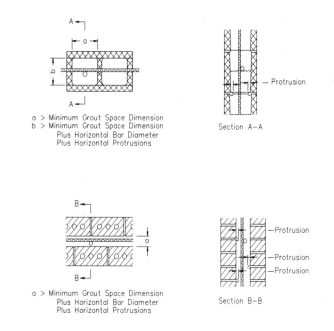

Fig. 3.1-1—Grout space requirements

wall construction made of differing material, this Code requires the strength of each wythe and of grouted collar joints to equal or exceed f'_m for the portion of the structure considered. If a multiwythe wall is designed as a composite wall, the compressive strength of each wythe or grouted collar joint shall equal or exceed f'_m.

CHAPTER 4—EMBEDDED ITEMS—ANCHORAGE OF MASONRY TO FRAMING AND EXISTING CONSTRUCTION

4.1—Embedded and conduits, pipes and sleeves

4.1.1 Conduits, pipes, and sleeves not harmful to mortar and grout can be embedded within the masonry, but the capacity of the wall shall not be less than that required by design. The effects of a reduction in section properties in the areas of pipe embedment should be considered. Horizontal pipes located in the planes of walls may affect the wall's load capacity.

For the integrity of the structure, all conduit and pipe fittings within the masonry should be carefully positioned and assembled. The coupling size should be considered when determining sleeve size.

Aluminum should not be used in masonry unless it is effectively coated or covered. Aluminum reacts with ions, may also react electrolytically with steel, causing cracking and/or spalling of the masonry. Aluminum electrical conduits present a special problem since stray electric current accelerates the adverse reaction.

Pipes and conduits placed in masonry, whether surrounded by mortar or grout or placed in unfilled spaces, need to allow unrestrained movement.

4.2—Anchorage of masonry to structural members, frames, and other construction

Masonry does not always behave in the same manner as its structural supports or adjacent construction. The designer should consider these differential movements and the forces resulting from their restraint. The type of connection chosen should transfer only the loads planned. While some connections transfer loads perpendicular to the wall, other devices transfer loads within the plane of the wall. Details shown in Fig. 4.2-1 are representative examples and allow movement within the plane of the wall. While load transfer usually involves masonry attached to structural elements such as beams or columns, the connection of nonstructural elements such as door and window frames should also be investigated.

4.3—Connectors

Connectors are of a variety of sizes, shapes and uses. In order to perform properly they should be identified on the Project Drawings.

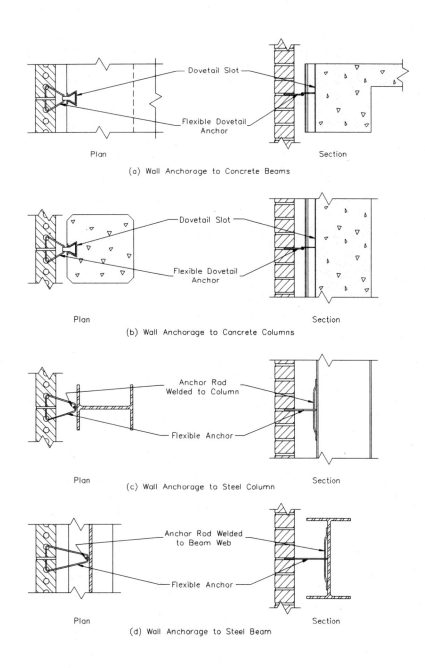

Fig 4.2-1—Wall anchorage details

PART 3—ANALYSIS AND DESIGN

CHAPTER 5—GENERAL ANALYSIS AND DESIGN REQUIREMENTS

5.1—Scope

The design procedures are "working stress" methods in which the stresses resulting from service loads do not exceed permissible service load stresses.

Linear elastic materials following the Hooke's Law are assumed, i.e., deformations (strains) are linearly proportional to the loads (stresses). All materials are assumed to be homogeneous and isotropic, and sections that are plane before bending remain plane after bending. These assumptions are adequate within the low range of working stresses under consideration. The allowable stresses are fractions of the specified compressive strength, resulting in conservative factors of safety.

Service load is the load which is assumed by the general building code to actually occur when the structure is in service. The stresses allowed under the action of service loads are limited to values within the elastic range of the materials.

Empirical design procedures of Chapter 9 are permitted in certain instances. Members not working integrally with the structure such as partition or panel walls, or any member not (or not permanently) absorbing or transmitting forces resulting from the behavior of the structure under loads may be designed empirically. A masonry shear wall would be an integral structural part while some wall partitions, because of their method of construction or attachment, would not. Empirical design is permitted for buildings of limited height and low seismic exposure.

5.2—Loading

The provisions establish design load requirements. If the service loads specified by the general building code differ from those of ASCE 7, the general building code governs. The Architect/Engineer may decide to use the more stringent requirements.

5.2.3 Lateral load stability must be provided by a braced structural system. Partitions, infill panels and similar elements, may not be a part of the lateral load resisting system if properly isolated. However, when they resist lateral forces due to their rigidity, they should be considered in analysis.

5.2.4 Service loads are not the sole source of stresses. The structure must also resist forces from the sources listed. The nature and extent of some of these forces may be greatly influenced by the choice of materials, structural connections, and geometric configuration.

5.3—Load combinations

5.3.1 The load combinations given are based on ASCE 7 and apply only if the general building code has none. Nine load combinations are to be considered and the structure designed to resist the maximum stresses resulting from the action of any load combination at any point of the structure.

This Code requires that when simultaneous loading is routinely expected, as in the case of dead and live loads, the structure must be designed to fully resist the combined action of the loads prescribed by the general building code.

5.3.2 Previous editions of building codes have customarily used a higher allowable stress when considering wind or earthquake in a structure. This increase has come under attack and there has been some confusion as to the rationale for permitting the increase. The committee recognizes this situation but has opted to continue to increase allowable stresses in the traditional manner until documentation is available to warrant a change (see reference 5.1).

5.4—Design strength

The structural adequacy of masonry construction requires that the compressive strength of masonry equal or exceed the specified strength. The specified compressive strength f'_m on which design is based for each part of the structure must be shown on the Project Drawings.

5.5—Material properties

5.5.1 Proper evaluation of the building material movement from all sources is an important element of masonry design. Brick and concrete masonry may behave quite differently under normal loading and weather conditions. The committee has extensively studied available research information in the development of these material properties. However, the Committee recognizes the need for further research on this subject. The designer is encouraged to review industry standards for further design information and movement joint location. Material properties can be determined by appropriate tests of the materials to be used.

5.5.2 *Elastic moduli*—Modulus of elasticity for masonry has traditionally been taken as 1000 f'_m in previous masonry codes. Research has indicated, however, that lower values may be more typical. A compilation of the available research has indicated a large variation in the relationship of elastic modulus versus compressive strength of masonry. However, variation in procedures between one research investigation and another may account for much of the indicated variation. Furthermore, the type of elastic moduli being reported (i.e., secant modulus, tangent modulus, chord modulus, etc.) is not always identified. The committee decided the most appropriate elastic modulus for working stress design purposes is the slope of the stress strain curve below a stress value of 0.33f'_m, the allowable flexural compressive stress. Data at the bottom of the stress strain curve may be questionable due to the "seating" effect of the specimen during the initial loading phase if measurements are made on the testing machine platens. The committee therefore decided that the most appropriate elastic modulus for design purposes is the chord modulus from a stress value of 5 to 33 percent of the compressive strength of masonry (see Fig. 5.5-1).

Using the elastic modulus defined above, the committee is still evaluating the relationship between elastic modulus and compressive strength. Code Tables 5.5.2.2 and 5.5.2.3 for clay and concrete masonry, respectively, are based on currently available information. These tables correlate elastic modulus to masonry unit strength rather than f'_m, which has been used previously. The specified compressive strength of masonry f'_m is not used because the actual compressive strength of masonry often significantly exceeds the specified value particularly for clay masonry.

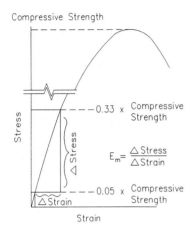

Fig 5.5-1—Chord modulus of elasticity of masonry

By relating elastic modulus of masonry to masonry unit strength, the contribution of each wythe to composite action is better taken into account in design calculations than would be the case if the elastic modulus of all parts of a composite wall were based on one specified compressive strength of masonry. The relationship between the modulus of rigidity and the modulus of elasticity has historically been given as $0.4E_m$. No experimental evidence exists to support this relationship.

5.5.3 *Thermal expansion coefficients*—Temperature changes cause material expansion and contraction. This material movement is theoretically reversible. These thermal expansion coefficients are slightly higher than mean values for the assemblage.[5.2,5.3,5.4]

Thermal expansion for concrete masonry[5.2,5.5] will vary with aggregate type.

5.5.4 *Moisture expansion coefficient of clay masonry*—Fired clay products expand upon contact with moisture, and the material does not return to its original size upon drying.[5.3,5.4] This is a long-term expansion as clay particles react with atmospheric moisture. Continued expansion has been reported for $7\frac{1}{2}$ years. Moisture expansion is reversible in concrete masonry.

5.5.5 *Shrinkage coefficients of concrete masonry*—Concrete masonry is a portland cement based material which will shrink due to moisture loss and carbonation.[5.5] Moisture- controlled units must be kept dry in order to retain the lower shrinkage values. The total linear drying shrinkage is determined by ASTM C 426. The shrinkage of clay masonry is negligible.

5.5.6 *Creep coefficients*—When continuously stressed, these materials gradually deform in the direction of stress

application. This movement is referred to as creep and is load and time dependent.[5.5,5.6] The values given are maximum values.

5.6—Deflection of beams and lintels

These deflection limits apply to beams of all materials which support unreinforced masonry.

These empirical requirements limit excessive deflections which may result in damage to the supported masonry. Where supported masonry is designed in accordance with Chapter 7 it is assumed that crack width in masonry will be controlled by the reinforcement so the deflection requirements are waived.

5.7—Lateral load distribution

The design assumptions for masonry buildings include the use of a braced structural system. The distribution of lateral loads to the members of the resisting structural system is a function of the rigidities of the structural system and of the horizontal diaphragms. The method of connection at intersecting walls and between walls and floor and roof diaphragms determines if the wall participates in the resisting structural system. Lateral loads from wind and seismic forces are normally considered to act in the direction of the principal axes of the structure. Lateral loads may cause forces in walls both perpendicular and parallel to the direction of the load. Horizontal torsion can be developed due to eccentricity of the applied load with respect to the center of rigidity.

The analysis of lateral load distribution should be in accordance with accepted engineering procedures. The analysis should rationally consider the effects of openings in shear walls and whether the masonry above the openings allows them to act as coupled shear walls. See Fig. 5.7-1. The interaction of coupled shear wall is complex and further information may be obtained from Reference 5.7.

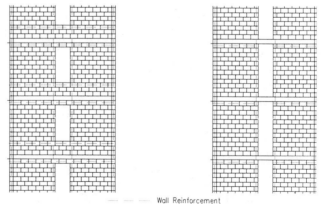

Note: Connecting elements between coupled shear walls typically require horizontal and vertical (not shown) reinforcement to transfer shear.

Elevation of Coupled Shear Wall   Elevation of Non-Coupled Shear

Fig. 5.7-1—Coupled and noncoupled shear walls

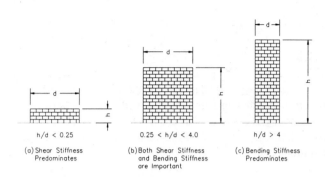

Fig 5.7-2—Shear wall stiffness

Computation of the stiffness of shear walls should consider shearing and flexural deformations. A guide for solid shear walls (i.e., with no openings) is given in Fig. 5.7-2. For nongrouted hollow unit shear walls, the use of equivalent solid thickness of wall in computing web stiffness is acceptable.

5.8—Multiwythe walls

5.8.2 *Composite action*—Multiwythe walls will act monolithically if sufficient shear strength is developed at the wythe interfaces. See Fig. 5.8-1. Shear transfer is achieved with headers crossing the collar joint or with mortar or grout filled collar joints. When mortar or grout filled collar joints are relied upon to transfer shear, wall ties are required to ensure structural integrity of the collar joint. Composite action requires that the stresses occurring at the interfaces are within the allowable limits prescribed.

Composite masonry walls generally consist of either brick-to-brick, block-to-block or brick-to-block wythes with the collar joint filled with mortar or grout, and the wythes connected with metal ties. The collar joint thickness ranges from ⅜ to 4 in. (9.5 to 102 mm). The joint may contain either vertical or horizontal reinforcement, or reinforcement may be placed in either the brick or block wythe. Composite walls are particularly advantageous for resisting high loads, both in-plane and out of plane.

Limited test data [5.8,5.9,5.28] are available to document shear strength of collar joints in masonry.

Test results [5.8,5.9] show that shear bond strength of collar joints could vary from as low as 5 psi (0.034 MPa) to as high as 100 psi (0.69 MPa) depending on type and condition of the interface, consolidation of the joint and type of loading. McCarthy et al.[5.8] reported an average value of 52 psi (0.36 MPa) with a coefficient of variation of 21.6 percent. A low bound allowable shear value of 5 psi (0.034 MPa) is considered to account for the expected high variability of the interface bond. With some units, Type S mortar slushed collar joints may have better shear bond characteristics than Type N mortar. Results show that thickness of joints, unit absorption and reinforcement have a negligible effect on shear bond strength. Grouted collar joints have higher allowable shear bond stress than the mortar

collar joints.[5.9] Requirements for masonry headers (Fig. 9.7-1) are empirical and taken from prior codes. The net area of the header should be used in calculating the stress even if a "solid" unit which allows up to 25 percent coring is used. Headers do not provide as much ductility as metal tied wythes with filled collar joints. The influence of differential movement is especially critical when headers are used. The committee does not encourage the use of headers.

A strength analysis has been demonstrated by Porter and Wolde-Tinsae[5.18,5.19] for composite walls subjected to combined in-plane shear and gravity loads. In addition, these authors have shown adequate behavioral characteristics for both brick-to-brick and brick-to-block composite walls with a grouted collar joint.[5.20,5.23] Finite element models for analyzing the interlaminar shearing stresses in collar joints of composite walls have been investigated by Anand et al.[5.24,5.27] They found that the shear stresses were principally transferred in the upper portion of the wall near the point of load application for the in-plane loads. Thus, below a certain distance the overall strength of the composite is controlled by the global strength of the wall, providing that the wythes are acting compositely.

The size, number, and spacing of wall ties, shown in Fig. 5.8-2, has been determined from past experience. The limitation of Z-ties to walls of other than hollow units is also based on past experience.

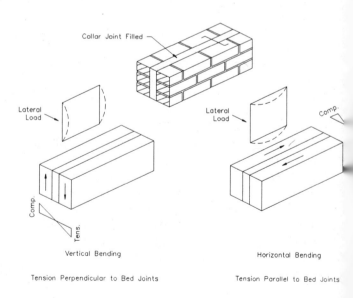

Fig 5.8-1—Stress distribution in multiwythe walls of compo. masonry

5.8.3 *Noncomposite action*—Multiwythe walls may be constructed so that each wythe is separated from the others by a space which may be crossed only by ties. The ties force compatible lateral deflection, but no composite action exists in the design. Weak axis bending moments caused by either gravity loads or lateral loads are assumed to be distributed to each wythe in proportion to its relative stiffness. See Fig. 5.8-3 for stress distribution innoncomposite walls. Loads due to supported horizontal members are to be carried by the wythe closest to center of span as a result of the deflection of the horizontal member.

The size, number and spacing of metal ties (Fig. 5.8-2) have been determined from past experience. In cavity walls, this Code limits the thickness of the cavity to $4^1/_2$ in. (114 mm) to assure adequate performance. If cavity width exceeds $4^1/_2$ in. (114 mm), the ties must be designed to carry the loads imposed upon them based on a rational analysis taking into account buckling, tension, pullout, and load distribution. The NCMA[5.10] and Canadian Standards Association, CSA[5.11] have recommendations for use in the design of ties for walls with wide cavities. The term cavity is used when the net thickness is 2 in. (51 mm) or greater. Two in. (51 mm) is considered the minimum space required for resistance to water penetration. A continuous air space of lesser thickness is referred to as a void (unfilled) collar joint. Requirementsfor adjustable ties are shown in Fig. 5.8-4. They are based on the results in Reference 5.12.

5.9—Columns

5.9.1 Columns are isolated members usually under axial compressive loads and flexure that, if damaged, may cause the collapse of other members; sometimes of an entire structure. These critical structural elements warrant the special requirements of this section that were selected after extensive committee consideration.

5.9.1.1 The minimum side dimension of 8 in. (203 mm) results from practical considerations.

5.9.1.2 The limit of 25 for the effective height-to-least nominal dimension ratio is based on experience. Data are currently lacking to justify a larger ratio. See Fig. 5.9-1 for effective height determination.

5.9.1.3 The minimum eccentricity of axial load (Fig. 5.9-2) results from construction imperfections not otherwise anticipated by analysis.

In the event that actual eccentricity exceeds the minimum eccentricity required by this Code, the actual eccentricity should be used. This Code requires that stresses be checked independently about each principal axis of the member (Fig. 5.9-2).

5.9.1.4 Minimum vertical reinforcement is required in masonry columns to prevent brittle collapse. The maximum percentage limit in column vertical reinforcement was established based on the committee's experience. Four bars are required so ties can be used to provide a confined core of masonry.

5.9.1.6 Lateral reinforcement in columns performs two functions. It provides the required support to prevent buckling for longitudinal column reinforcing bars acting in compression and provides resistance to diagonal tension for

columns acting in shear.[5.13] Ties may be located in the mortar joint.

The requirements of this Code are modeled on those for reinforced concrete columns. Except for permitting #2 ties outside of Seismic Performance Category D or E they reflect all applicable provisions of the reinforced concrete code.

5.10—Pilasters

Pilasters are masonry members which can serve one of several purposes. They may be visible, projecting from one or both sides of the wall, or hidden within the thickness of the wall as shown in Fig. 5.10-1. Pilasters aid in the lateral load resistance of masonry walls and may carry vertical loads.

5.11—Load transfer at horizontal connections

Masonry walls, pilasters, and columns may be connected to horizontal elements of the structure, and may rely on the latter for lateral support and stability. The mechanism through which the interconnecting forces are transmitted may involve bond, mechanical anchorage, friction, bearing, or a combination thereof. The designer must assure that, regardless of the type of connection, the interacting forces are safely resisted.

In flexible frame construction, the relative movement (drift) between floors may generate forces within the members and the connections. This Code requires the effects of these movements to be considered in design.

5.12—Concentrated loads

5.12.1 Masonry laid in running bond will distribute the axial compressive stress resulting from a concentrated load along the length of wall as described in this Code. Stress can only be transmitted across the head joints of masonry laid in running bond. Thus when other than running bond is used, concentrated loads can only be spread across the length of one unit unless a bond beam or other technique is used to distribute the load (Fig. 5.12-1).

5.12.2 When the supporting masonry area is larger on all sides than the bearing area, this Code allows distribution of concentrated loads over a bearing area A_2, determined as illustrated in Fig. 5.12-2. This is permissible because the confinement of the bearing area by surrounding masonry increases the bearing capacity of the wall in the vicinity of concentrated loads.

5.13—Section properties

5.13.1 *Stress computations*—Minimum net section is often difficult to establish in hollow unit masonry. The designer may choose to use the minimum thickness of the face shells of the units as the minimum net section. The minimum net section may not be the same in the vertical and horizontal directions.

For masonry of hollow units, the minimum cross-sectional area in both directions may conservatively be based on the minimum face shell thickness.[5-29]

Solid clay masonry units are permitted to have coring up to a maximum of 25 percent of their gross cross- sectional area.

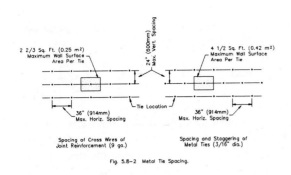

Fig 5.8-2—Wall tie spacing for multiwythe walls

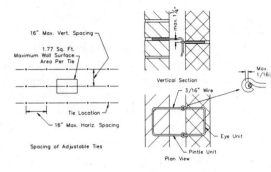

Fig. 5.8-4—Adjustable ties

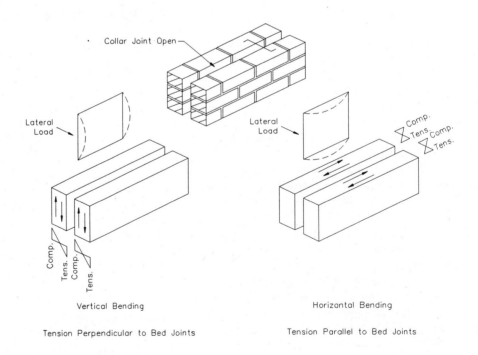

Fig. 5.8-3—Stress distribution in multiwythe walls of noncomposite masonry

If data (see Section 1.3) shows that there is reliable restraint against translation and rotation at the supports the "effective height" may be taken as low as the distance between points of inflection for the loading case under consideration

Fig. 5.9-1—Effective height, h, of column, wall or pilaster, in.

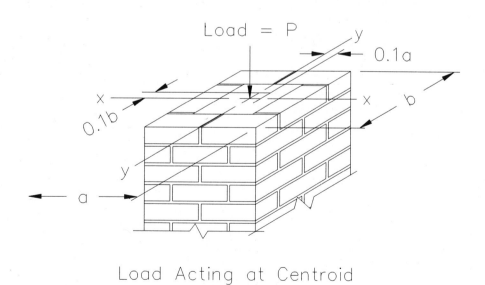

Load Acting at Centroid

Fig. 5.9-2—Minimum design eccentricity in columns

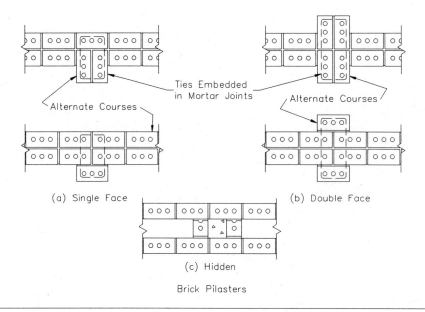

(a) Single Face   (b) Double Face

(c) Hidden

Brick Pilasters

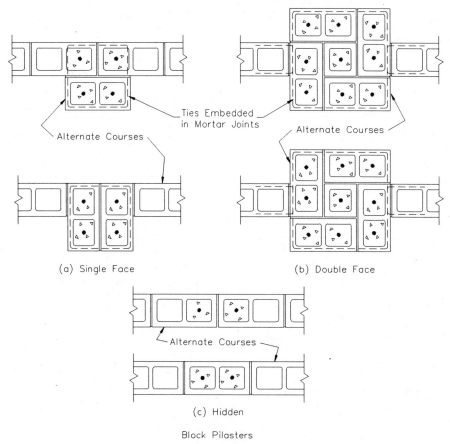

(a) Single Face   (b) Double Face

(c) Hidden

Block Pilasters

Fig. 5.10-1—Typical pilaster

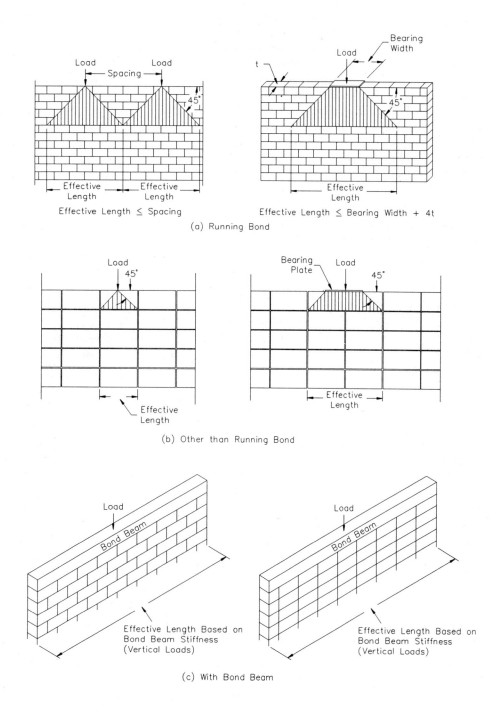

Fig. 5.12-1—Load distribution

For such units the net cross-sectional area may be taken as equal to the gross cross-sectional area, except as provided in 5.8.2.2.(c) for masonry headers. Several conditions of net area are shown in Fig. 5.13-1.

Since the elastic properties of the materials used in members designed for composite action differ, equal strains produce different levels of stresses in the components. To compute these stresses, a convenient transformed section with respect to the axis of resistance is considered. The resulting stresses developed in each fiber are related to the actual stresses by the ratio E_1/E_x between the moduli of elasticity of the weakest material in the member and of the materials in the fiber considered. Thus, to obtain the transformed section, fibers of the actual section are conceptually widened by the ratio E_x/E_1. Stresses computed based on the section properties of the transformed section, with respect to the axis of resistance considered, are then multiplied by E_x/E_1 to obtain actual stresses.

5.13.2 *Stiffness*—Stiffness is a function of the extent of cracking. However, this Code equations for design in Chapter 6 are based on the member's uncracked moment of inertia. Also since the extent of tension cracking in shear walls is not known in advance, this Code allows the determination of stiffness to be based on uncracked section properties. For reinforced masonry the stiffness calculations based on the cracked section will yield more accurate results.

The section properties of masonry members may vary from point to point. For example, in a single wythe concrete masonry wall made of hollow ungrouted units, the cross-sectional area will vary through the unit height. Also, the distribution of material varies along the length of the wall or unit. For stiffness computations, an average value of the appropriate section property, i.e., cross-sectional area and/or moment of inertia, is considered adequate for design. The average net cross-sectional area of the member would in turn be based on average net cross-sectional area values of the masonry units and the mortar joints composing the member.

5.13.3 *Radius of gyration*—The radius of gyration is the square root of the ratio of bending moment of inertia to cross-sectional area. Since stiffness is based on the average net cross-sectional area of the member considered, this same area should be used in the computation of radius of gyration.

5.13.4 *Intersecting walls*—Connections of webs to flanges of shear walls may be accomplished by running bond, metal anchors or bond beams. Achieving stress transfer at a T intersection with running bond only is difficult. A running bond connection should be as shown in Fig. 5.13-2 with a "T" geometry over their common intersection.

An alternate method making use of metal strap anchors is shown in Fig. 5.13-3. Bond beams shown in Fig. 5.13-4 are the third means of connecting webs to flanges.

When the flanges are connected at the intersection, be sure to included them in the design. The effective width of the flange is an empirical requirement. The effective flange width is shown in Fig. 5.13-5.

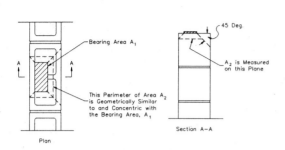

Fig. 5.12-2—Bearing areas

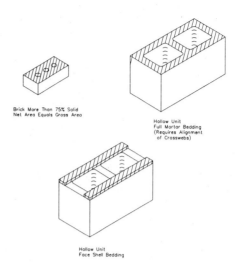

Fig. 5.13-1—Net cross-sectional areas

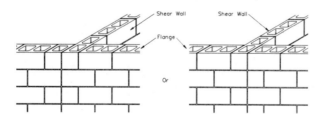

Fig. 5.13-2—Running bond lap at intersection

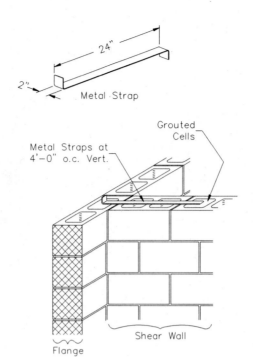

Fig. 5.13-3—Metal straps and grouting at wall intersections

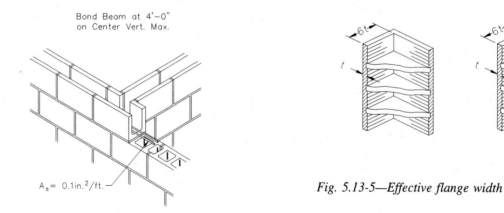

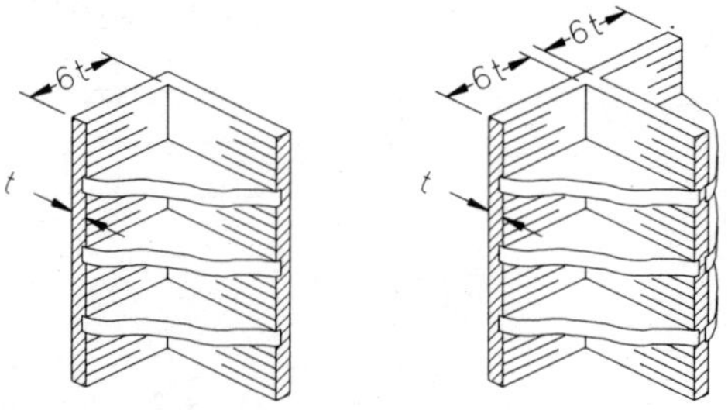

Fig. 5.13-5—Effective flange width

Fig. 5.13-4—Bonding ties and grouting for flanged shear walls

5.14—Anchor bolts solidly grouted in masonry

5.14.1 *Test design requirements*—The design of anchor bolts is based on physical testing. Testing may be used to establish higher working loads than those calculated by Section 5.14.2. Many types of anchor bolts; expansion anchors, toggle bolts, sleeve anchors, etc., are not included in Section 5.14.2 and therefore such anchors must be designed using test data. ASTM E 448 requires only three tests. The variation in test results for anchors embedded in masonry warrants an increase to the minimum of five stipulated. The variability of anchor bolt strength in masonry and the possibility that anchor bolts may be used in a nonredundant manner results in a safety factor of five.

5.14.2 *Plate, headed, and bent bar anchor bolts*—These design values apply only to the specific bolts mentioned. They are readily available and are depicted in Fig. 5.14-1.

5.14.2.1 The minimum embedment depth requirement is considered a practical minimum based on typical construction methods for embedding bolts in masonry. The validity of allowable shear and tension equations for small embedment depths, less than four bolt diameters, has not been verified by tests.

5.14.2.2 The results of tests on anchor bolts in tension showed that anchors failed by pullout of a conically shaped section of masonry, or by failure of the anchor itself. Bent bar anchor bolts (J-bolts) often failed by completely sliding out of the specimen. This was due to straightening of the bent end. Eq. (5-1) is the allowable tension load based on masonry failure. The area A_p is the projected area of the assumed failure cone. The cone originates at the bearing point of the embedment and radiates at 45° in the direction of the pull (See Fig. 5.14-2). Comparisons of Eq. (5-1) to test results obtained by Whitlock[5.15] show an average factor of safety of approximately eight. Eq. (5-2) is the allowable load for anchor bolts based on failure of the bolt.

The equation allows one-fifth of the yield load for all types of anchor bolts. Eq. (5-1) and (5-2) are plotted in Fig. 5.14-3.

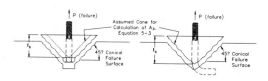

Fig. 5.14-2—Tension failure of masonry

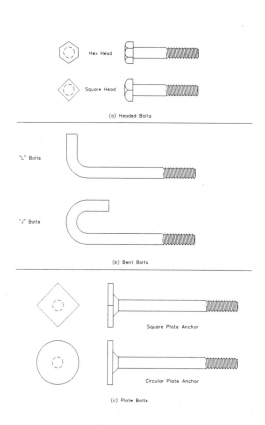

Fig. 5.14—1—Anchor bolts

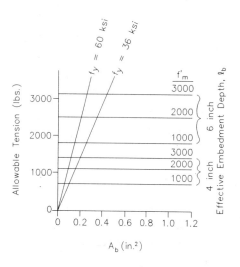

Fig. 5.14-3—Allowable axial tension on anchor bolts

As anchor bolts are spaced closer together, the stresses within the masonry begin to become additive. Therefore, where the spacing between the anchors is less than $2\ell_b$, this Code requires that the projected areas used to calculate allowable load be reduced to reflect the additive stresses in the area of cone overlap as shown in Fig. 5.14-4.

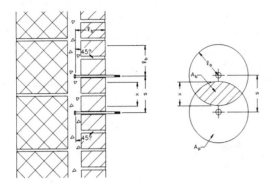

Fig. 5.14-4—Anchor bolt cone area overlap

Test results[5.15] have shown that the pullout strength of bent bar anchors correlated best with a reduced embedment depth. This may be explained with reference to Fig. 5.14-5. Due to the radius of the bend, stresses are concentrated at a point closer than the full embedment distance.

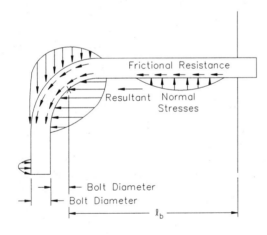

Fig. 5.14-5—Stress distribution on bent anchor bars

5.14.2.3 *Allowable loads in shear*—Eq. (5-5) was derived from research done by Hatzinikolas et al.,[5.16] and, when compared to tests done by Hatzinikolas and Whitlock,[5.15] the factors of safety range from approximately six to eight, respectively. Eq. (5-6) is based on the "shear friction" concept with a coefficient of friction equal to 0.6 and a safety factor of five. Fig. 5.14-6 contains plots of Eq. (5-5) and (5-6).

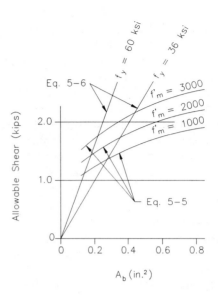

Fig. 5.14-6—Allowable shear on anchor bolts

Sufficient edge distances must be provided such that failures do not occur in modes that are not accounted for in the design equations.

(a) The reason is that with this amount of edge distance, a full failure cone can develop.

(b) The edge distance in the direction of the shear load was derived by equating the following expression:

$$V = \sqrt[4]{f'_m}(\pi\, m^2/2) \quad \text{(one-half stress cone directed toward free edge)}$$

and

$$V = 0.6\,(\pi D^2/4)f_y \quad \text{(anchor steel strength)}$$

This resulted in the following expression:

$$m = D\sqrt{(0.6/8)}\;(f_y/f'_m)$$

For f_y = 60,000 psi (414 MPa) and f'_m = 1,000 psi (7.89 MPa), the required edge distance, m, equals $12D$.

5.14.2.4 *Combined shear and tension*—Test results[5.15] have shown that the strength of anchor bolts follows a circular interaction line. However, for simplicity and additional conservatism, this Code requires a straight line interaction between allowable shear and tension loads.

5.15—Framed construction
Exterior masonry walls connected to structural frames are used primarily as non-bearing curtain walls. Regardless of the structural system used for support, there are differential movements between the structure and the wall. These differential movements may occur separately or in combination and may be due to the following:

1) Temperature increase or decrease of either the structural frame or the masonry wall,

2) Moisture and freezing expansion of brick or shrinkage of concrete block walls.

3) Elastic shortening of columns from axial loads shrinkage or creep.

4) Deflection of supporting beams.

5) Sideways in multiple-story buildings.

6) Foundation movement.

Since the tensile strength of masonry is low, these differential movements must be accommodated by sufficient clearance between the frame and masonry and flexible or slip-type connections.

Structural frames should not be infilled with clay masonry to increase resistance to inplane lateral forces without considering the differential movements listed above.

Wood, steel, or concrete columns may be surrounded by masonry serving as a decorative element. Masonry walls may be subject to forces as a result of their interaction with other structural components. Since the masonry element is often much stiffer, the load will be carried first by the masonry. These forces, if transmitted to the surrounding masonry, should not exceed the allowable stresses of the masonry used. Alternately, there should be sufficient clearance between the frame and masonry. Flexible ties should be used to allow for the deformations.

Beams or trusses supporting masonry walls are essentially embedded, and their deflections should be limited to the allowable deflections for the masonry being supported. See Section 5.6 for requirements.

5.16—Stack bond masonry

The requirements separating running bond from stack bond are shown in Fig. 5.16-1. The amount of steel required in

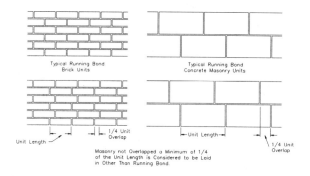

Fig. 5.16-1—Running bond masonry

this section is an arbitrary amount to provide continuity across the head joints. This reinforcement can be used to resist load.

References

5.1. Ellifrit, D.S., "The Mysterious $^{1}/_{3}$ Stress Increase," *Engineering Journal*, ASIC, 4th Quarter, 1977.

5.2. Copeland, R.E., "Shrinkage and Temperature Stresses in Masonry," ACI JOURNAL, *Proceedings* V. 53, No. 8, American Concrete Institute, Detroit MI, Feb. 1957, pp. 769-780.

5.3. Plummer, Harry C., *Brick and Tile Engineering*, Brick Institute of America, Reston, VA, 1962, 736 pp.

5.4. Grimm, C.T., "Probabilistic Design of Expansion Joints in Brick Cladding," *Proceedings*, 4th Canadian Masonry Symposium, University of Fredericton, 1986, V. 1, pp. 553-568.

5.5. Kalouseb, L., "Relation of Shrinkage to Moisture Content in Concrete Masonry Units," *Paper* No. 25, Housing and Home Finance Agency, Washington, D.C., 1954.

5.6. Lenczner, D. and Salahuddin, J., "Creep and Moisture Movements in Masonry Piers and Walls," *Proceedings*, 1st Canadian Masonry Symposium, University of Calgary, June 1976, pp. 72-86.

5.7. *Structural Design of Tall Concrete and Masonry Buildings*, Monograph on Planning and Design of Tall Buildings, V. CB, Council on Tall Buildings and Urban Habitat/American Society of Civil Engineers, New York, NY, 1978, 960 pp.

5.8. McCarthy, J.A., Brown, R.H., and Cousins, T.E., "An Experimental Study of the Shear Strength of Collar Joints in Grouted and Slushed Composite Masonry Walls," *Proceedings*, 3rd North American Masonry Conference, University of Texas, Arlington, TX, June 1985, pp. 39-1—39-16.

5.9. Williams, R. and Geschwinder, L., "Shear Stress Across Collar Joints in Composite Masonry," presented at *Proceedings*, 2nd North American Masonry Conference, University of Maryland, College Park, MD, *Paper* No. 8.

5.10. "Joint Reinforcement and Other Metal Ties for Concrete Masonry Walls," *NCMA-TEK Bulletin* No. 64, National Concrete Masonry Association, Herndon, VA, 1975, 4 pp.

5.11. "Connectors for Masonry," (CAN 3-A370-M84), Canadian Standards Association, Rexdale, Ontario, 1984.

5.12. "Development of Adjustable Wall Ties," *ARF Project* No. B869, Illinois Institute of Technology, Chicago, IL, Mar. 1963.

5.13. Pfister, James. F., "Influence of Ties on the Behavior of Reinforced Concrete Columns," ACI JOURNAL, *Proceedings* V. 61, No. 5, American Concrete Institute, Detroit, MI, May 1964, pp. 521-537.

5.14. Kolodziejski, E.A. and Drysdale, R.G., "Shear Stress of Intersecting Masonry Walls," *Report* No. 29, Faculty of Engineering, McMasters University, Hamilton, Ontario, Dec. 1982.

5.15. Brown, R.H. and Whitlock, A.R., "Strength of Anchor Bolts in Concrete Masonry," *Journal of the Structural Division*, American Society of Civil Engineers, New York, NY, Vol. 109, No. 6, June, 1983, pp. 1362-1374.

5.16. Hatzinikolos, M., Longworth, J., and Warwaruk, J., "Strength and Behavior of Anchor Bolts Embedded in Concrete Masonry," *Proceedings*, 2nd Canadian Masonry Conference, Carleton University, Ottawa, Ontario, June, 1980. pp. 549-563.

5.17. Hamid, A.A., Ziab, G., and Nawawy, Ed, "Modulus of Elasticity of Concrete Block Masonry," *Proceeding* 4th North American Masonry Conference, University of

California, Los Angeles, CA, Aug. 1987, pp. 7-1—7-13.

5.18.   Porter, M.L., Wolde-Tinsae, A.M., and Ahmed, M.H., "Strength Analysis of Composite Walls," *Advances in Analysis of Structural Masonry*, Proceedings of Structures Congress '86, American Society of Civil Engineers, New York, NY, 1986.

5.19.   Porter, M.L., Wolde-Tinsae, A.M., and Ahmed, M.H., "Strength Design Method for Brick Composite Walls," *Proceedings*, 4th International Masonry Conference, London, Aug. 1987.

5.20.   Wolde-Tinsae, A.M., Porter, M.L., and Ahmed, M.H., "Shear Strength of Composite Brick-to-Brick Panels," *Proceedings*, 3rd North American Masonry Conference,University of Texas, Arlington, TX, June 1985, pp. 40-1—40-13.

5.21.   Wolde-Tinsae, A.M., Porter, M.L., and Ahmed, M.H., "Behavior of Composite Brick Walls," *Proceedings*, 7th International Brick Masonry Conference, Melbourne, New South Wales, Feb. 1985, V. 2, pp. 877-888.

5.22.   Ahmed, M.H., Porter, M.L., and Wolde-Tinsae, A.M., "Behavior of Reinforced Brick-to-Block Walls," PhD dissertation, M. H. Ahmed, Iowa State University, Ames, IA, 1983, Part 2A.

5.23.   Ahmed, M.H., Porter, M.L., and Wolde-Tinsae, A.M., "Behavior of Reinforced Brick-to-Block Walls," PhD dissertation, M. H. Ahmed, Iowa State University, Ames, 1983, Part 2B.

5.24.   Anand, Subhash C. and Young, David. T., "A Finite Element Model for Composite Masonry," *Proceedings*, American Society of Civil Engineers, V. 108, ST12, New York, NY, Dec. 1982, pp. 2637-2651.

5.25.   Anand, S.C., "Shear Stresses in Composite Masonry Walls," *New Analysis Techniques for Structural Masonry*, American Society of Civil Engineers, New York, NY, Sept. 1985, pp. 106-127.

5.26.   Anand, S.C. and Stevens, D.J., "Shear Stresses in Composite Masonry Walls Using a 2-D Modes," *Proceedings*, 3rd North American Masonry Conference, University of Texas, Arlington, TX, June 1985, p. 41.

5.27.   Anand, S.C. and Rahman, M.A., "Temperature and Creep Stresses in Composite Masonry Walls," *Advances in Analysis of Structural Masonry*, American Society of Civil Engineers, New York, NY, 1986, pp. 111-133.

5.28.   Colville, J., Matty, S.A., and Wolde-Tinsae, A.M., "Shear Capacity of Mortared Collar Joints," *Proceedings*, 4th North American Masonry Conference, University of California, Los Angeles, CA, Aug. 1987, V. 2 pp. 60-1—60-15.

5.29.   "Section Properties for Concrete Masonry," *NCMA-TEK* 14-1, National Concrete Masonry Association, Herndon, VA, 1990.

CHAPTER 6—UNREINFORCED MASONRY

6.1—Scope

This chapter provides for the design of masonry members in which tensile stresses, not exceeding allowable limits, are resisted by the masonry. This has previously been referred to as unreinforced or plain masonry. Flexural tensile stresses may result from bending moments from eccentric vertical loads or lateral loads.

A fundamental premise is that under the effects of design loads, masonry remains uncracked. One must be aware, however, that stresses due to restraint against differential movement, temperatures change, moisture expansion and shrinkage combine with the design load stresses. Stresses due to restraint should be controlled by joints or other construction techniques to insure that the combined stresses do not exceed the allowable.

6.2—Stresses in reinforcement

Reinforcement may be placed in masonry walls to control the effects of movements from temperature changes or shrinkage.

6.3—Axial compression and flexure

6.3.1—For a member solely subjected to axial load, the resulting compressive stress f_a should not exceed the allowable compressive stress F_a; in other words, f_a/F_a should not exceed 1. Similarly, in a member subjected solely to bending, the resulting compressive stress f_b in the extreme compression fiber should not exceed the allowable compressive stress F_b, or again, f_b/F_b should not exceed 1.

This Code requires that under combined axial and flexure loads, the sum of the quotients of the resulting compression stresses to the allowable $f_a/F_a + f_b/F_b$ does not exceed 1. This unity interaction equation is a simple portioning of the available allowable stresses to the applied loads, and is used to design masonry for compressive stresses. The unity formula can be extended when biaxial bending is present by replacing the bending stress quotients with the quotients of the calculated bending stress over the allowable bending stress for both axes.

In this interaction equation, secondary bending effects resulting from the axial load are ignored. A more accurate equation would include the use of a moment magnifier applied to the flexure term, f_b/F_b. Although avoidance of a moment magnifier term will produce nonconservative results, the committee decided not to include this term in Eq. (6-1) for the following reasons:

- At larger h/r values where moment magnification is more critical, the allowable axial load on the member will be limited by Code Eq. (6-2).
- For the practical range of h/r values, errors induced by ignoring the moment magnifier will be relatively small, less than 15 percent.
- The overall safety factor of 4 included in the allowable stress equations is sufficiently large to allow this simplification in the design procedure.

The requirement of Eq. (6-2) that the axial compressive load P not exceed $1/4$ of the buckling load P_e replaces the arbitrary upper limits on slenderness used in ACI 531.[6.16]

The purpose of Eq. (6-2) is to safeguard against a premature stability failure caused by eccentrically applied axial load. The equation is not intended to be used to check adequacy for combined axial compression and flexure. Therefore, in Eq. (6-6), the value of the eccentricity "e" that is to be used to calculate P_e is the actual eccentricity of the applied compressive load. The value of "e" is not to be calculated as M_{max} divided by P where M_{max} is a moment caused by other than eccentric load.

Eq. (6-2) is an essential check since the allowable compressive stress, for members with an h/r ratio in excess of 99, has been developed assuming only a nominal eccentricity of the compressive load. Thus, when the eccentricity of the compressive load exceeds the minimum eccentricity of $0.1t$, Eq. (6-4) will overestimate the allowable compressive stress and Eq. (6-2) may control.

The allowable stress values for F_a, presented in Eqs. (6-3) and (6-4) are based on an analysis of the results of axial load tests performed on clay and concrete masonry elements. A fit of an empirical curve to this test data, Fig. 6.3-1, indicates that members having an h/r ratio not exceeding 99, fail under loads below the Euler Buckling Load at a stress level equal to

$$f_m' [1 - (h/140r)^2]$$

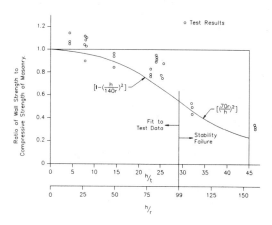

Fig. 6.3-1—Slenderness effects on axial compressive strength

Thus, for members having an h/r ratio not exceeding 99, this Code allows axial load stresses not exceeding $1/4$ of the aforementioned failure stress.

Applying the Euler theory of buckling to members having resistance in compression but not in tension, references 6-1, 6-2 and 6-3 show that for a solid section, the critical compressive load for these members can be expressed by the formula

$$P_e = (\pi^2 EI/h^2)(1 - 2e/t)^3$$

in which

I = uncracked moment of inertia

e = eccentricity of axial compressive load with respect to the member longitudinal centroidal axis.

In the derivation of this buckling load equation, tension cracking is assumed to occur prior to failure.

For h/r values in excess of 99, the limited test data is approximated by the buckling load.

For a solid rectangular section $r = \sqrt{t^2/12}$. Making this substitution into the buckling load equation gives

$$P_e = (\pi^2 EI/h^2)[1 - 0.577(e/r)]^3 \qquad (6\text{-}6)$$

Transforming the buckling equation using a minimum eccentricity of $0.1t$ (from Section 5.9.1.3) and an elastic modulus equal to $1000 f'_m$, the axial compressive stress at buckling failure amounts approximately to $[70(r/h)]^2 f'_m$. Thus, for members having an h/r ratio in excess of 99, this Code allows an axial load compressive stress not exceeding $\frac{1}{4}$ of this failure stress [Eq. (6-4)].

Flexure tests of masonry to failure have shown[6.4,6.5,6.6,6.7] that the compressive stress at failure computed by the straight line theory exceeds that of masonry failing under axial load. This phenomenon is attributed to the restraining effect of less highly strained compressive fibers on the fibers of maximum compressive strain. This effect is less pronounced in hollow masonry than solid masonry; however, the test data indicate that, computed by the straight line theory, the compressive stress at failure in hollow masonry subjected to flexure exceeds by $\frac{1}{3}$ that of the masonry under axial load. Thus, to maintain a factor of safety of 4 in design, the committee considered it conservative to establish the allowable compressive stress in flexure as

$$f_b = \frac{4}{3} \cdot (\frac{1}{4})f'_m = (\frac{1}{3})f'_m$$

6.3.1.1 Allowable flexural tensile stresses for portland cement lime mortar are traditional values.

For masonry cement and air entrained portland cement lime mortar there are no conclusive research data and hence flexural tensile stresses are based on existing requirements in other codes.

The tensile stresses listed are only for tension due to flexure under out-of-plane loading. Flexural tensile stresses can be offset by axial compressive stress, but the resultant tensile stress due to combined bending and axial compression cannot exceed the allowable flexural tensile stress. Note, no values for allowable tensile stress are given in this Code for in-plane bending because flexural tension in walls should be carried by reinforcement from in-plane bending.

Variables affecting tensile bond strength of brick masonry normal to bed joints include mortar properties, unit initial rate of absorption, surface condition, workmanship and curing condition. For tension parallel to bed joints the strength and geometry of the units will also have an effect on tensile strength.

Test data using a bond wrench[6.8,6.9] revealed tensile bond strength normal to bed joints ranging from 30 psi (0.21 MPa) to 190 psi (1.3 MPa). This wide range is attributed to the multitude of parameters affecting tensile bond strength.

Test results[6.9,6.17] show that masonry cement mortars and mortars with high air content generally have lower bond

strength than portland cement-lime mortars.

Tests conducted by Hamid[6.10] show the significant effect of the aspect ratio (height to least dimension) of the brick unit on the flexural tensile strength. The increase in the aspect ratio of the unit results in an increase in strength parallel to bed joints and a decrease in strength normal to bed joints.

Research work[6.11] on flexural strength of concrete masonry has shown that grouting has a significant effect in increasing strength capacity over ungrouted masonry. A three-fold increase in tensile strength normal to bed joints was achieved using fine grout as compared to ungrouted masonry. The results also show that, within a practical range of strength, the actual strength of grout is not of major importance. For tension parallel to bed joints, a 133 percent increase in flexural strength was achieved by grouting all cells. Grout cores change the failure mode from stepped-wise cracking along the bed and head joints for hollow walls to a straight line path along the head joints and unit for grouted walls.

For partial grouting, the footnote permits interpolation between the fully grouted value and the hollow unit value based on the percentage of grouting. A concrete masonry wall with Type S portland cement-lime mortar grouted 50 percent and stressed normal to the bed joints would have an allowable stress midway between 68 psi (0.47 MPa) and 25 psi (0.17 MPa), hence an allowable stress of 46.5 psi (0.32 MPa).

6.4—Axial tension

Tensile stresses in masonry walls due to axially applied load are not permitted. If axial tension develops in walls due to uplift of connected roofs or floors, the walls must be reinforced to resist the tension. Cumulative compressive stress from dead load can be used to offset axial tension. Masonry columns are required to have vertical reinforcing by Section 5.9.1.4.

6.5—Shear

Three modes of shear failure in unreinforced masonry are possible:

(a) Diagonal tension cracks form through the mortar and masonry units.

(b) Sliding occurs along a straight crack at horizontal bed joints.

(c) Stepped cracks form, alternating from head joint to bed joint.

In the absence of suitable research data the committee recommends that the allowable shear stress values given in Code Section 6.5.2 be used for limiting out-of-plane shear stresses.

6.5.1 The theoretical parabolic stress distribution is used to calculate shear stress rather than the average stress. Many other codes use average shear stress so direct comparison of allowable values is not valid. Effective area requirements are given in Section 5.13.1. For rectangular sections this equates to $\frac{3}{2} \cdot V/A$. This equation is also used to calculate shear stresses for composite action.

6.5.2 Shear stress allowable values are applicable to shear walls without reinforcement. The values given are based on recent research.[6.12-6.15] The 0.45 coefficient of friction, increased from 0.20, is shown in these tests. N_v is normally based on dead load.

6.5.3 Shear stress at these locations have not normally been included in previous codes. See the commentary for Section 5.8.2.

References

6.1. Colville, J., "Simplified Design of Load Bearing Masonry Walls," *Proceedings*, 5th International Symposium on Loadbearing Brickwork, *Publication* No. 27, British Ceramic Society, London, Dec. 1978, pp. 2171- 2234.

6.2. Colville, J., "Stress Reduction Design Factors for Masonry Walls," *Proceedings*, American Society of Civile Engineers, V. 105, ST10, New York, NY, Oct. 1979, pp. 2035-2051.

6.3. Yokel, Felix Y., "Stability and Load Capacity of Members with no Tensile Strength," *Proceedings*, American Society of Civil Engineers, V. 97, ST7, New York, NY, July 1971, pp. 1913-1926.

6.4. Hatzinikolas, M., Longworth, J., and Marworuk, J., "Concrete Masonry Walls," *Structural Engineering Report* No. 70, Department of Civil Engineering, University of Alberta, Canada, Sept. 1978.

6.5. Fattal, S.G. and Cattaneo, L.E., "Structural Performance of Masonry Walls Under Compression and Flexure," *Building Science Series* No. 73, National Bureau of Standards, Washington, D.C., 1976, 57 pp.

6.6. Yokel, Felix Y., and Dikkers, Robert D., "Strength of Load-Bearing Masonry Walls," *Proceedings*, American Society of Engineers, V. 97, ST5, New York, NY, `May 1971, pp. 1593-1609.

6.7. Yokel, Felix Y., and Dikkers, Robert D., Closure to "Strength of Load-Bearing Masonry Walls," *Proceedings*, American Society of Engineers, V. 99, ST5, New York, NY, May 1973, pp. 948-950.

6.8. Brown, R. and Palm, B., "Flexural Strength of Brick Masonry Using the Bond Wrench," *Proceedings*, 2nd North American Masonry Conference, University of Maryland, College Park, MD, Aug. 1982.

6.9. Hamid, A.A., "Bond Characteristics of Sand-Molded Brick Masonry," *The Masonry Society Journal*, V. 4, No. 1, Boulder, CO, Jan.-June 1985, pp. T-18—T-22.

6.10. Hamid, A.A., "Effect of Aspect Ratio of the Unit on the Flexural Tensile Strength of Brick Masonry," *The Masonry Society Journal*, Boulder, CO, V. 1, Jan.-June 1981.

6.11. Drysdale, R.G. and Hamid, A.A., "Effect of Grouting on the Flexural Tensile Strength of Concrete Block Masonry," *The Masonry Society Journal*, V. 3, No. 2, Boulder, CO, July-Dec. 1984, pp. T-1—T-9.

6.12. Woodward, K. and Ranking, F., "Influence of Vertical Compressive Stress on Shear Resistance of Concrete Block Masonry Walls," U.S. Department of Commerce, National Bureau of Standards, Washington, D.C., Oct. 1984, 62 pp.

6.13. Pook, L.L., Stylianou, M.A., and Dawe, J.L., "Experimental Investigation of the Influence of Compression on the Shear Strength of Masonry Joints," *Proceedings*, 4th Canadian Masonry Symposium, Fredericton, New Brunswick, June 1986, pp. 1053-1062.

6.14. Nuss, L.K., Noland, J.L., and Chinn, J., "The Parameters Influencing Shear Strength Between Clay Masonry Units and Mortar," *Proceedings*, North American Masonry Conference, University of Colorado, Boulder, CO, Aug. 1978.

6.15. Hamid, Ahmad A., Drysdale, Robert G., and Heidebrecht, Arthur C., "Shear Strength of Concrete Masonry Joints," *Proceedings*, American Society of Engineers, V. 105, ST7, New York, NY, July 1979, pp. 1227-1240.

6.16. ACI Committee 531, "Building Code Requirements for Concrete Masonry Structures (ACI 531-79) (Revised 1983)," American Concrete Institute, Detroit, MI, 1983, 20 pp.

6.17. Ribar, J., "Water Permeance of Masonry: A Laboratory Study," *Masonry: Properties and Performance*, STP-778, ASTM, Philadelphia, PA, 1982.

6.18. "Research Data and Comments in Support of: Recommended Building Code Requirements for Engineered Concrete Masonry," National Concrete Masonry Association, Herndon, VA, unpublished.

6.19. "Recommended Practice for Engineered Brick Masonry," Brick Institute of America, Reston, VA, 1980, 337 pp.

CHAPTER 7—REINFORCED MASONRY

7.1—Scope

The requirements covered in this chapter pertain to the design of masonry previously referred to as "reinforced masonry." The term, reinforced masonry, has been avoided to more accurately describe the conditions of design covered in Chapters 6 and 7. Additionally it will avoid confusion with masonry designed in accordance with the provisions of Chapter 6 in which the effect of joint and other reinforcement used in construction is neglected in the design.

Tension still develops in the masonry, but it is not considered to be effective in resisting design loads.

7.2—Steel reinforcement

7.2.1 *Allowable stresses*—These values have been in use for many years.

7.3—Axial compression and flexure

See Commentary for 6.3 and 6.3.1.

7.3.2 *Allowable forces and stresses*—This Code limits the compressive stress in masonry members based on the type of load acting on the member. The compressive force at the section resulting from axial loads or from the axial component of combined loads is calculated separately, and is limited to the values permitted in Section 7.3.2.1. Equation 7.1 or 7.2 will control the capacity of columns with large axial loads. The coefficient of 0.25 provides a factor of safety of about 4.0 against crushing of masonry. The coefficient of 0.65 was determined from tests of reinforced masonry columns and is taken from previous masonry codes[7.1, 7.2]. A second compressive stress calculation must be performed considering the combined effects of the axial load component and flexure at the section and shall be limited to the values permitted in Section 7.3.2.2. (See commentary for Section 6.3.)

7.3.2.2 See commentary for Section 6.3.1 for information on F_b.

The interaction equation used in Section 6.3 is not applicable for reinforced masonry and is therefore not included in this chapter.

7.3.3 *Effective compressive width per bar*—The effective width of the compressive area for each reinforcing bar must be established. Fig. 7.3-1 depicts the limits for the conditions stated. Limited research[7.3] is available on this subject.

The limited ability of head joints to transfer stress when the masonry is laid in stack bond is recognized by the requirements for bond beams. Masonry units with open ends that are solidly grouted will transfer stress as indicated in Section 6.5.2(d) and can qualify as running bond.

The center-to-center bar spacing maximum is a limit to keep from overlapping areas of compressive stress. The 72 in. (1829 mm) maximum is an empirical choice of the committee.

7.3.4 *Beams*—The requirements for masonry members outlined are relatively straight forward, and follow generally accepted engineering practice.

The minimum bearing length of 4 in. (102 mm) in the direction of span is considered a reasonable minimum for masonry beams over door and window openings to prevent concentrated compressive stresses at the edge of the opening. This requirement should also apply to beams and lintels in the plane of the wall.

To minimize lateral torsional buckling, Section 7.3.3.4 requires lateral bracing of the compression face in accordance with standard limits for beams of other materials. The requirement applies to simply supported beams as written. With continuous or fixed beams the spacing may be increased.

7.5—Shear

To compensate for a simplified method of analysis, and unknowns in construction, the shear stresses allowed by this Code are conservative. When reinforcement is added to masonry, the shear resistance of the element is increased. Priestley and Bridgemen[7.4] concluded from a series of tests that shear reinforcement is effective in providing resistance only if it is designed to carry the full shear load. Thus, most codes do not add the shear resistance provided by the masonry to that provided by the steel. The amount of design shear reinforcement is specified to resist one hundred percent of the applied shearing load. See Commentary Section 6.5 and the flow chart for design of masonry members resisting shear shown in Fig. 7.5.1.

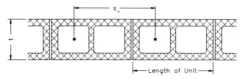

For Other Than Running Bond Masonry with Bond Beam Spaced Less Than or Equal to 48 in. and Running Bond Masonry, b Equals the Lessor of:

b = s
b = 6t
b = 72 in.

For Masonry in Other Than Running Bond with Bond Beams Spaced Greater Than 48 in., b Equals the Lessor of:

b = s
b = Length of Unit

Fig. 7.3-1—Width of compression area

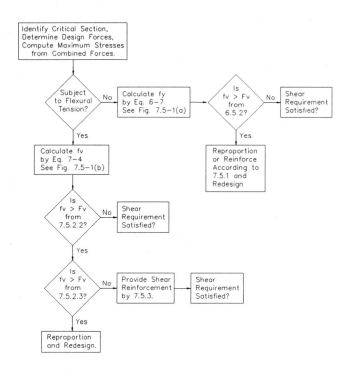

Fig. 7.5-1—Flow chart for shear design

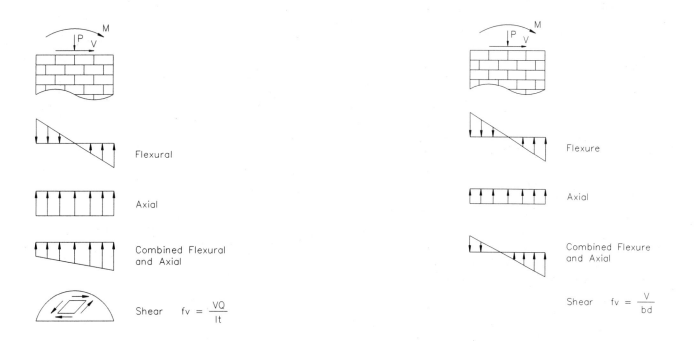

7.5-1(a)—Illustration of design section which is subjected to tension

Fig. 7.5-1(b)—Illustration of design section which is subjected to tension

7.5.2 Eq. (7-4) through (7-10) in Code Section 7.5.2 are derived from previous masonry codes.[7.1,7.5,7.6]

7.5.2.1 Shear forces can act both vertically and horizontally under wind and seismic conditions in shear walls. Since the beams are reinforced and will exhibit flexural cracking, the classical shear stress calculation used in Chapter 6 is replaced with an approximation of the maximum shear stress below the neutral axis. The approximation results from deleting the term "j" in the equation $f_v = V/bjd$.

7.5.2.3(a) The limits on the calculated shear stress in beams are in conformance with those found in previous masonry codes.

7.5.3 Eq. (7-10) may be derived by assuming a 45 deg shear crack extended from the extreme compression fiber to the centroid of the tension steel which is the distance d. Forces are summed in the direction of the shear reinforcement and the doweling resistance of the longitudinal reinforcement is neglected. In Code Eq. (7-10), for shear walls without shear reinforcement, for shear parallel to the plane of the wall, d_v may be substituted for d. Notice that for such shear walls d_v may be either horizontal or vertical, depending on the direction of the shear and resulting reinforcement.

For shear walls, the longitudinal reinforcement is normally vertical and distributed along the length of the wall. The shear reinforcement is normally horizontal. In the development of the equation for shear walls, the 45 deg crack extends through more horizontal reinforcement than that obtained by using the depth to the centroid of the steel d. Thus, the use of d_v is justified. However, the designer must be cautioned that this is not always the case. For example, in a 10 ft (3.05 m) shear wall with vertical reinforcement located 2 ft (0.6 m) from each end (with no other vertical reinforcement) it would be unconservative to use d_v and the maximum reinforced length may be used in place of d_v.

7.5.3.1 The assumed shear crack is at 45 deg to the longitudinal reinforcement. Thus, a maximum spacing of $d/2$ is specified to assure that each crack is crossed by at least one bar. The 48 in. (1219 mm) maximum spacing is an arbitrary choice which has been in codes for many years.

7.5.4 Shear across collar joints in composite masonry walls is transferred by the mortar or grout in the collar joint. Shear stress in the collar joint or at the interface between the wythe and the collar joint is limited to the allowable stresses in Section 5.8.1.2. Shear transfer by wall ties or other reinforcement across the collar joint is not considered.

7.5.5 The beam or wall loading within $d/2$ of the support is assumed to be carried in direct compression or tension to the support without increasing the shear load, provided no concentrated load occurs within the $d/2$ distance.

References

7.1. ACI Committee 531, "Building Code Requirements for Concrete Masonry Structures (ACI 531-79) (Revised 1983)," American Concrete Institute, Detroit, MI, 1983, 20 pp.

7.2 "Recommended Practices for Engineered Brick Masonry,"Brick Institute of America, Reston, VA, pp. 337.

7.3. Dickey, W. and MacIntosh, A., "Results of Variation of b' or Effective Width in Flexure in Concrete Block Panels," Masonry Institute of America, Los Angeles, CA, 1971.

7.4. Priestley, M.J.N., and Bridgeman, D.O., "Seismic Resistance of Brick Masonry Walls," *Bulletin*, New Zealand National Society for Earthquake Engineering (Wellington), V. 7, No. 4, Dec. 1974, pp. 167-187.

7.5. "Specification for the Design and Construction of Load Bearing Concrete Masonry," (TR-75B), National Concrete Masonry Association, Herndon, VA, 1979.

7.6. "Building Code Requirements for Engineered Brick Masonry," Brick Institute of America, Reston, VA, 1969, 36 pp.

CHAPTER 8—DETAILS OF REINFORCEMENT

8.1—Scope

In setting the provisions of this chapter, the committee used the ACI 318 Code[8.1] as a guide. Some of the requirements were simplified and others dropped, depending on their suitability for application to masonry. Formulae relative to embedment and splicing have been simplified due to the use of a safety factor larger for masonry than for reinforced concrete.

8.2—Size of reinforcement

8.2.1   Limits on size of reinforcement are based on accepted practice and successful performance in construction. The #11 limit is arbitrary, but research 7.3 shows that distributed small bars provide better performance than fewer large bars. Properties of reinforcement are given in Table 8.2.1.

8.2.2   Adequate flow of grout for the achievement of good bond is achieved with this limitation. It also limits the size of reinforcement when combined with Section 3.1.2.

Table 8.2.1—Physical properties of steel reinforcing wire and bars

Designation		Diameter, in. (mm)	Area, in.² (mm²)	Perimeter, in. (mm)
Wire				
W1.1 (11 gage)		0.121 (3.07)	0.011 (7.10)	0.380 (9.65)
W1.7 (9 gage)		0.148 (3.76)	0.017 (11.0)	0.465 (11.8)
W2.1 (8 gage)		0.162 (4.12)	0.020 (12.9)	0.509 (12.9)
W2.8 (3/16 wire)		0.187 (4.75)	0.027 (17.4)	0.587 (14.9)
W4.9 (¹/₄ wre)		0.250 (6.35)	0.049 (31.6)	0.785 (19.9)
Bars	**Metric**			
#3		0.375 (9.53)	0.11 (71.0)	1.178 (29.92)
	10	0.445 (11.3)	0.16 (100)	1.398 (35.5)
#4		0.500 (12.7)	0.20 (129)	1.571 (39.90)
#5	15	0.625 (15.9)	0.31 (200)	1.963 (49.86)
#6		0.750 (19.1)	0.44 (284)	2.456 (62.38)
	20	0.768 (19.5)	0.47 (300)	2.413 (61.3)
#7		0.875 (22.2)	0.60 (387)	2.749 (69.83)
	25	0.992 (25.2)	0.76 (500)	3.118 (79.2)
#8		1.000 (25.4)	0.79 (510)	3.142 (79.81)
#9		1.128 (28.7)	1.00 (645)	3.544 (90.02)
	30	1.177 (29.9)	1.09 (700)	3.697 (93.9)
#10		1.270 (32.2)	1.27 (819)	3.990 (101.3)
	35	1.406 (35.7)	1.55 (1000)	4.417 (112.2)
#11		1.410 (35.8)	1.56 (1006)	4.430 (112.5)

8.2.3   The function of joint reinforcement is to control the size and spacing of cracks caused by volume changes in masonry as well as to resist tension.[8.2] Joint reinforcement is commonly used in concrete masonry to minimize shrinkage cracking. The restriction on wire size ensures adequate performance. The maximum wire size of one-half the joint thickness allows free flow of mortar around joint reinforcement. Thus, a ³/₁₆ in. (4.8 mm) diameter wire can be placed in a ³/₈ in. (9.5 mm) joint.

8.3—Placement limits for reinforcement

Placement limits for reinforcement are based on successful construction practice over many years. The limits are intended to facilitate the flow of grout between bars. A minimum spacing between bars in a layer prevents longitudinal splitting of the masonry in the plane of the bars. Use of bundled bars in masonry construction is rarely required. Two bars per bundle is considered a practical maximum. It is important that bars be placed accurately. Reinforcing bar positioners are available to control bar position.

8.4—Protection for reinforcement

8.4.1   Reinforcing bars are traditionally not galvanized. The masonry cover retards corrosion of the steel. Cover is measured from the exterior masonry surface to the outer-most surface of the steel to which the cover requirement applies. It is measured to the outer edge of stirrups, or ties, if transverse reinforcement encloses main bars. Masonry cover includes the thickness of masonry units, mortar, and grout. At bed joints the protection for reinforcement is the total thickness of mortar and grout from the exterior of the mortar joint surface to outer-most surface of the steel.

The condition "masonry face exposed to earth or weather" refers to direct exposure to moisture changes (alternate wetting and drying) and not just temperature changes.

8.4.2   Since masonry cover protection for joint reinforcement is minimal, the protection of joint reinforcement in masonry is required in accordance with the standard specification.

8.4.3   Corrosion resistance requirements are included since masonry cover varies considerably for these items. The exception for anchor bolts is based on current industry practice.

8.5—Development of reinforcement embedded in grout

8.5.1   From a point of peak stress in reinforcement, some length of reinforcement or anchorage is necessary through which to develop the stress. This development length or anchorage is necessary on both sides of such peak stress points, on one side to transfer stress into and on the other to transfer stress out of the reinforcement. Often the reinforcement continues for a considerable distance on one side of a critical stress point so that calculations need involve only the other side, e.g., the negative moment reinforcement continuing through a support to the middle of the next span.

All bars and longitudinal wires must be deformed.

8.5.2   *Embedment of bars and wires in tension*—Eq. (8-1) can be derived from the basic development length expression and an allowable bond stress u for deformed bars in grout of 160 psi.[8.3,8.4] Recent research [8.9] has shown that epoxy-coated reinforcing bars require longer development length than uncoated reinforcing bars. The 50 percent increase in development length is consistent with ACI 318 provisions.

$$l_d = d_b F_s / 4u = d_b F_s / 4(160) = 0.0015 d_b F_s$$

8.5.3   *Embedment of flexural reinforcement*—Fig. 8.5-1 illustrates the embedment requirements of flexural reinforcement in a typical continuous beam. Fig. 8.5-2 illustrates the embedment requirements in a typical continuous wall that is not part of the lateral load resisting system.

8.5.3.1(b) Critical sections for a typical continuous beam are indicated with a "c" or an "x" in Fig. 8.5-1. Critical sections for a typical continuous wall are indicated with a "c" in Fig. 8.5-2.

8.5.3.1(c) The moment diagrams customarily used in design are approximate. Some shifting of the location of maximum moments may occur due to changes in loading, settlement of supports, lateral loads, or other causes. A diagonal tension crack in a flexural member without stirrups may shift the location of the calculated tensile stress approximately a distance d toward a point of zero moment. When stirrups are provided, this effect is less severe, although still present.

To provide for shifts in the location of maximum moments, this Code requires the extension of reinforcement a distance d or $12d_b$ beyond the point at which it is theoretically no longer required to resist flexure, except as noted.

Cutoff points of bars to meet this requirement are illustrated in Fig. 8.5-1.

When bars of different sizes are used, the extension should be in accordance with the diameter of bar being terminated. A bar bent to the far face of a beam and continued there may logically be considered effective, in satisfying this section, to the point where the bar crosses the middepth of the member.

8.5.3.1(d) Peak stresses exist in the remaining bars wherever adjacent bars are cut off, or bent, in tension regions. In Fig. 8.5-1 an "x" mark is used to indicate the peak stress points remaining in continuing bars after part of the bars have been cut off. If bars are cut off as short as the moment diagrams allow, these stresses become the full F_s, which requires a full embedment length as indicated. This

extension may exceed the length required for flexure.

8.5.3.1(e) Evidence of reduced shear strength and loss of ductility when bars are cut off in a tension zone has been

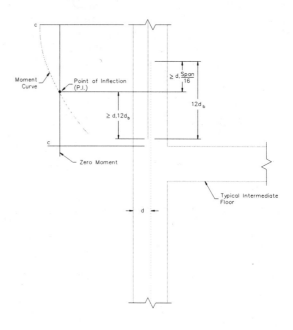

Fig. 8.5-2—Development of flexural reinforcement in a typical wall

reported in Reference 8.5. As a result, this Code does not permit flexural reinforcement to be terminated in a tension zone unless special conditions are satisfied. Flexure cracks tend to open early wherever any reinforcement is terminated in a tension zone. If the stress in the continuing reinforcement and the shear strength are each near their limiting values, diagonal tension cracking tends to develop prematurely from these flexure cracks. Diagonal cracks are less likely to form where shear stress is low. A lower steel stress reduces the probability of such diagonal cracking.

8.5.3.1(f) Members, such as corbels, members of variable depth arches and others where steel stress f_s does not decrease linearly in proportion to a decreasing moment require special consideration for proper development of the flexural reinforcement.

8.5.3.2 *Development of positive moment reinforcement*—When a flexural member is part of a primary lateral load resisting system, loads greater than those anticipated in design may cause reversal of moment at supports. As a consequence some positive reinforcement is required to be anchored into the support. This anchorage assures ductility of response in the event of serious overstress, such as from blast or earthquake. The use of more reinforcement at lower stresses is not sufficient. The full anchorage requirement does not apply to any excess reinforcement provided at the support.

8.5.3.3 *Development of negative moment reinforcement*—Negative reinforcement must be properly

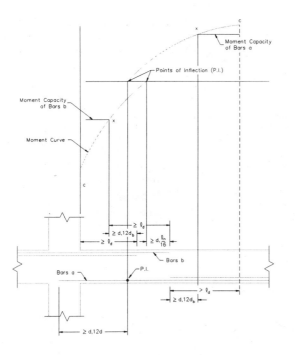

Fig. 8.5-1—Development of flexural reinforcement in a typical continuous beam

anchored beyond the support faces by extending the reinforcement l_d into the support. Other methods of anchoring include the use of a standard hook or suitable mechanical device.

Section 8.5.3.3(b) provides for possible shifting of the moment diagram at a point of inflection, as discussed under Commentary Section 8.5.3.1(c). This requirement may exceed that of Section 8.5.3.1(c) and the more restrictive governs.

8.5.4 *Standard hooks*—Standard hooks are shown in Fig. 8.5-3.

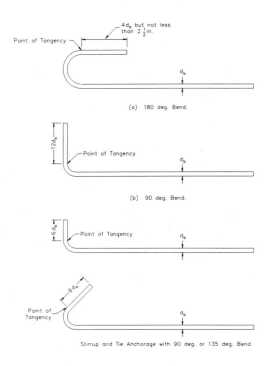

Fig. 8.5-3—Standard hooks

8.5.5 *Minimum bend diameter for reinforcing bars*—Standard bends in reinforcing bars are described in terms of the inside diameter of bend since this is easier to measure than the radius of bend.

A broad survey of bending practices, a study of ASTM bend test requirements, and a pilot study of and experience with bending Grade 60 bars were considered in establishing the minimum diameter of bend. The primary consideration was feasibility of bending without breakage. Experience since has established that these minimum bend diameters are satisfactory for general use without detrimental crushing of grout.

8.5.5.2 The allowable stress developed by a standard hook, 7500 psi (51.7 MPa), is the accepted permissible value in masonry design. Substituting this value into Eq. (8-1) yields the equivalent embedment length given. This value is less than half that given in Reference 8.1.

8.5.5.3 In compression, hooks are ineffective and cannot be used as anchorage.

8.5.6 *Development of shear reinforcement*

8.5.6.1(a) Stirrups must be carried as close to the compression face of the member as possible because near ultimate load the flexural tension cracks penetrate deeply.

8.5.6.1(b) The requirements for anchorage of U-stirrups for deformed reinforcing bars and deformed wire are illustrated in Fig. 8.5-4.

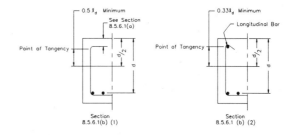

Fig. 8.5-4—Anchorage of U-stirrups (deformed reinforcing bars and deformed wire)

8.5.6.1(b)(1) When a standard hook is used, 0.5 l_d must be provided between $d/2$ and the point of tangency of the hook.

This provision may require a reduction in size and spacing of web reinforcement, or an increase in the effective depth of the beam, for web reinforcement to be fully effective.

8.5.6.1(c) and (e) U-stirrups which enclose a longitudinal bar obviously have sufficient resistance in the tension zone of the masonry.

8.5.6.2 *Welded wire fabric*—Although not often used in masonry construction, welded wire fabric provides a convenient means of placing reinforcement in a filled collar joint. See Reference 8.7 for more information.

8.5.7 *Splices of reinforcement*—The importance of continuity in the reinforcement through proper splices is emphasized by the different requirements for the stress level to be transferred in the various types of splices.[8.8]

8.5.7.1 *Lap splices*—Perhaps the easiest splices to achieve, the length of the splice is based on the allowable stress in the reinforcement.

8.5.7.1(a) The length of lap splices is greater than the required development length of the bars, indicating the assumption of a lower bond stress at the splice.

8.5.7.1(b) If individual bars in noncontact lap splices are too widely spaced, an unreinforced section is created, forcing a potential crack to follow a zigzag line. Lap splices may occur with the bars in adjacent grouted cells if these requirements are met.

8.5.7.2 *Welded splices*—A full welded splice is primarily intended for large bars (#6 and larger) in main members. The tensile strength requirement of 125 percent of specified yield strength will ensure sound welding, adequate also for compression. It is desirable that splices be capable of developing the ultimate tensile strength of the bars spliced, but practical limitations make this ideal condition difficult to attain. The maximum reinforcement stress used in design under this Code is based upon yield strength. To ensure sufficient strength in splices so that brittle failure can be avoided, the 25 percent increase above the specified yield strength was selected as both an adequate minimum for safety and a practicable maximum for economy.

8.5.7.3 *Mechanical connections*—Full mechanical connections are also required to develop 125 percent of the yield strength, in tension or compression as required, for the same reasons discussed for full welded splices.

8.5.7.4 *End bearing splices*—Experience with end bearing splices has been almost exclusively with vertical bars in columns. If bars are significantly inclined from the vertical, special attention is required to insure that adequate end bearing contact can be achieved and maintained. The lateral tie requirements prevent end bearing splices from sliding.

References

8.1. ACI Committee 318, "Building Code Requirements for Reinforced Concrete (ACI 318-83)," American Concrete Institute, Detroit, MI 1983, 111 pp.

8.2. Dickey, W.L., "Joint Reinforcement and Masonry," *Proceedings*, 2nd North American Masonry Conference, University of Maryland, College Park, MD Aug. 1982.

8.3. Gallagher, E.F., "Bond Between Reinforcing Steel and Brick Masonry," *Brick and Clay Record*, V. 5, Cahners Publishing Co., Chicago, IL Mar. 1935, pp. 86-87.

8.4. Richart, F.E., "Bond Tests Between Steel and Mortar," Structural Clay Products Institute (Brick Institute of America), Reston, VA, 1949.

8.5. Ferguson, Phil M., and Matloob, Farid N., "Effect of Bar Cutoff on Bond and Shear Strength of Reinforced Concrete Beams," ACI JOURNAL, *Proceedings* V. 56, No. 1, July 1959, pp. 5-24.

8.6. ACI Committee 408, "Bond Stress the State-of-the-Art," ACI JOURNAL, *Proceedings* V. 63, No. 11, Detroit, MI, Nov. 1966, pp. 1161-1190.

8.7. Joint PCI/WRI Ad Hoc Committee on Welded Wire Fabric for Shear Reinforcement, "Welded Wire Fabric for Shear Reinforcement," *Journal*, Prestressed Concrete Institute, V. 25, No. 4, Chicago, IL, July-Aug. 1980, pp. 32-36.

8.8. ACI Committee 318, "Commentary on Building Code Requirements for Reinforced Concrete (ACI 318-83)," American Concrete Institute, Detroit, MI, 1983, 155 pp.

8.9. Treece, Robert A.,"Bond Strength of Epoxy-Coated Reinforcing Bars", Masters Thesis, Department of Civil Engineering, University of Texas at Austin, Austin, TX May, 1987.

CHAPTER 9—EMPIRICAL DESIGN OF MASONRY

9.1—Scope

Empirical rules and formulas for the design of masonry structures were developed by experience. These are part of the legacy of masonry's long use, predating engineering analysis. Design is based on the condition that gravity loads are reasonably centered on the bearing walls and the effect of any steel reinforcement, if used, is neglected. The masonry should be laid in running bond. Specific limitations on building height, seismic, wind and horizontal loads exist. Buildings are of limited height. Members not participating in the lateral force resisting system of a building may be empirically designed even though the lateral force resisting system is designed under Chapters 5, 6, 7, and 8.

These procedures have been compiled through the years.[9.1-9.5] The most recent of these documents[9.5] is the basis for this chapter.

Empirical design is a procedure of sizing and proportioning masonry elements, it is not design analysis. This procedure is conservative for most masonry construction. Empirical design of masonry was developed for buildings of smaller scale, with more masonry interior walls and stiffer floor systems than built today. Thus, the limits imposed are valid.

9.3—Lateral stability

Lateral stability requirements are a key provision of empirical design. Obviously shear walls must be in two directions to provide stability. Bearing walls can serve as shear walls. See Fig. 9.3-1 for cumulative length of shear walls. A minimum shear wall length equal to the story height should be used.

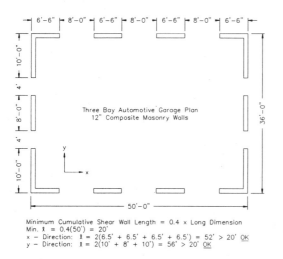

Minimum Cumulative Shear Wall Length = 0.4 x Long Dimension
Min. ℓ = 0.4(50') = 20'
x – Direction: ℓ = 2(6.5' + 6.5' + 6.5' + 6.5') = 52' > 20' OK
y – Direction: ℓ = 2(10' + 8' + 10') = 56' > 20' OK

Fig. 9.3-1—Cumulative length of shear walls

9.4—Compressive stress requirements

These are average compressive stresses, based on gross area using actual dimensions. The following conditions should be used as guidelines when concentrated loads are placed on masonry:

• For concentrated loads acting on the full wall thickness,

the allowable stresses under the load may be increased by 25 percent.

• For concentrated loads acting on concentrically placed bearing plates greater than one-half but less than full area, the allowable stress under the bearing plate may be increased by 50 percent.

The course immediately under the point of bearing should be a solid unit or filled solid with mortar or grout.

9.5—Lateral support

Lateral support requirements are included to limit the flexural tensile stress due to out-of-plane loads. Masonry headers resist shear stress and permit the entire cross-section to perform as a single element. This is not the case for non-composite walls connected with wall ties. For such non-composite walls, the use of the sum of the thicknesses of the wythes has been used successfully for a long time, and is a traditional approach that is acceptable within the limits imposed by Table 9.5.1.

9.6—Thickness of masonry

9.6.1 Experience of the committee has shown that the present ANSI A 41.1[9.4,9.5] thickness ratios are not always conservative. These requirements represent the consensus of the committee for more conservative design.

9.6.3 Foundation walls—Empirical criteria for masonry foundation wall thickness related to the depth of unbalanced fill have been contained in building codes and federal government standards for many years. The use of Table 9.6.3.1., which lists the traditional allowable backfill depths, is limited by a number of requirements which were not specified in previous codes and standards. These restrictions are enumerated in Section 9.6.3.1. Further precautions are recommended to guard against allowing heavy earth-moving or other equipment near enough to the foundation wall to develop high earth pressures. Experience with local conditions should be used to modify the values in Table 9.6.3.1 when appropriate.

9.7—Bond

Fig. 9.7-1 depicts the requirements listed.

9.8—Anchorage

The requirements of Sections 9.8.2.2 through 9.8.2.5 are less stringent than those of Section 5.13.4.2(e).

9.9—Miscellaneous requirements

9.9.4 *Corbelling*—The provision for corbelling up to one-half of the wall or unit thickness is valid only if the opposite side of the wall remains in its same plane. See Fig. 9.9-1 for maximum unit projection.

References

9.1. Baker, Ira Osborn, *A Treatise on Masonry Construction*, University of Illinois, Champaign, IL, 1889, 1899, 1903. Also, 10th Edition, John Wiley & Sons, New York, NY, 1909, 745 pp.

9.2. "Recommended Minimum Requirements for Masonry Wall Construction," *Publication* No. BH6, National Bureau of Standards, Washington, D.C., 1924.

9.3. "Modifications in Recommended Minimum Requirements for Masonry Wall Construction," National Bureau of Standards, Washington, D.C., 1931.

9.4. "American Standard Building Code Requirements for Masonry," (ASA A 41.1), American Standards Association, New York, NY, 1944.

9.5. "American Standard Building Code Requirements for Masonry," (ANSI A 41.1), American National Standards Institute, New York, NY, 1953 (1970).

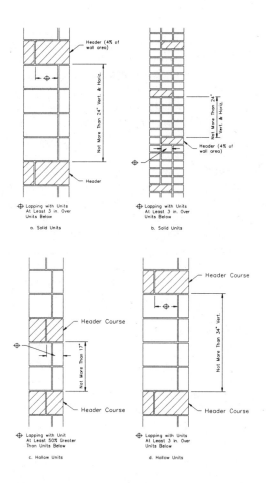

Fig. 9.7-1—Cross section of wall elevations

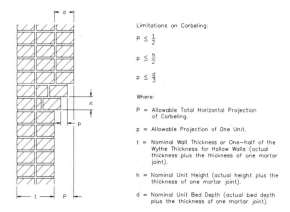

Fig. 9.9-1—Limitations on corbelling

CHAPTER 10—SEISMIC DESIGN REQUIREMENTS

10.1—Scope

The requirements in this chapter have been devised to improve performance of masonry construction when subjected to earthquake loads. ASCE 7 has been cited here as the appropriate reference for the distribution of seismic forces in order to avoid confusion in the event that the general building code has no provisions or is inconsistent with the type of distribution upon which these provisions are based.

The special provisions are presented in a cumulative format. Thus the provisions for Seismic Performance Category E include provisions for Seismic Performance Category D which include provisions for Seismic Performance Category C, and so on.

Seismic requirements for masonry veneers are found in Chapter 12, Veneers.

10.2—General

By reference to Section 5.1, the designer is permitted to use allowable stress design methods for reinforced masonry, allowable stress design for unreinforced masonry or empirical design. The alternate method in Section 10.2.2 permits a strength design methodology in which allowable stress values are modified to approximate strength value levels. The designer should note that the limitations of the Seismic Performance Categories may further limit the available design options. For instance, empirical design procedures are not permitted to be used for structures in Seismic Performance Categories D and E. Chapter 9 Empirical Design of Masonry does not permit empirical design for the lateral force-resisting system in Seismic Performance Categories B and C.

If the general building code has adopted the seismic load provisions of ASCE 7 the "strength" design procedures of this section should be used. If the general building code has seismic load provisions specifically intended for working stress design, the allowable stress design procedures of Section 5.1 should be used.

10.2.2 The strength of members and connections is based on working stress procedures modified by a factor. The nominal capacity is approximated as the allowable stress increased by $^1/_3$ (for the load combinations that include wind or earthquake in accordance with Section 5.3.2) and further multiplied by a factor of 2.5.

10.2.2.2 The resulting nominal strength is approximately 3.3 times the allowable value obtained by using allowable stress design methodology. The design strength is equal to the nominal strength times the strength reduction factor, ϕ, to achieve a reliable design level value.

Because of the modifications of allowable stress values to strength design levels, some element strengths are calculated using steel stresses in excess of the specified yield. This procedure is correct, and produces designs which are intended to give similar levels of performance as using working stresses in combination with service-level seismic loads.

10.3—Seismic Performance Category A

The general requirements of this Code provide for adequate performance of masonry construction in areas of low seismic risk.

10.4—Seismic Performance Category B

Although masonry may be designed by the provisions of Chapter 6, Unreinforced Masonry, Chapter 7, Reinforced Masonry, or Chapter 9, Empirical Design of Masonry, the lateral force-resisting system for structures in Seismic Performance Category B must be designed based on a structural analysis in accordance with Chapter 6 or 7. The provisions of Chapter 9 do not apply to the design of the lateral force-resisting system of buildings in Seismic Performance Category B.

10.5—Seismic Performance Category C

In addition to the requirements of Category B, minimum levels of reinforcement and detailing are required. The minimum provisions for improved performance of masonry construction in Seismic Performance Category C must be met regardless of the method of design.

10.5.3.1 Experience has demonstrated that one of the chief causes of failure of masonry construction during earthquakes is inadequate anchorage of masonry walls to floors and roofs. For this reason, an arbitrary minimum anchorage based upon previously established practice has been set. When anchorage is between masonry walls and wood framed floors or roofs, the designer should avoid the use of wood ledgers in cross-grain bending.

10.5.3.2 Experience has demonstrated that connections of structural members to masonry columns are vulnerable to damage during earthquakes unless properly anchored. Requirements are adapted from previously established practice developed as a result of the 1971 San Fernando earthquake.

10.5.3.3 The provisions of this section require an arbitrary minimum amount of reinforcement to be included in masonry wall construction. Tests reported in Reference 10.2 have confirmed that masonry construction reinforced as indicated performs adequately at this seismic load level. This minimum required reinforcement may also be used to resist design loads.

10.6—Seismic Performance Category D

10.6.3 The minimum amount of wall reinforcement has been a long-standing, standard empirical requirement in areas of high seismic loading. It is expressed as a percentage of gross cross-sectional area of the wall. It is intended to improve the ductile behavior of the wall under earthquake loading and assist in crack control. Since the minimum required reinforcement may be used to satisfy design requirements, at least $^1/_3$ of the minimum amount is reserved for the lesser stressed direction in order to ensure an appropriate distribution in both directions. The intent of Section 10.6.3.1 is to provide a minimum level of in-plane shear reinforcement to improve ductility.

10.6.4 Adequate lateral restraint is important for column reinforcement subjected to overturning forces due to earthquakes. Many column failures during earthquakes have been attributed to inadequate lateral tying. For this reason, closer spacing of ties than might otherwise be required is prudent. An arbitrary minimum spacing has been established through experience. Columns not involved in the lateral force resisting system should also be more heavily tied at the tops and bottoms for more ductile performance and better resistance to shear.

10.7—Seismic Performance Category E

10.7.1 and **10.7.2** See commentary Sections 10.5.3.3 and 10.6.3. The ratio of minimum horizontal reinforcement is increased to reflect the possibility of higher seismic loads. Where solidly grouted, open end hollow units are used part of the need for horizontal reinforcement is satisfied by the mechanical continuity provided by the grout core.

References

10.1. "Recommended Lateral Force Requirements," Seismology Committee, Structural Engineers Association of California, Sacramento, CA, Oct. 1986.

10.2. Gulkan, P., Mayes, R.L., and Clough, R.W., "Shaking Table Study of Single-Story Masonry Houses Volumes 1 and 2," *Report* No. UCB/EERC-79/23 and 24, Earthquake Engineering Research Center, University of California, Berkeley, CA, Sept. 1979.

10.3. "Tall Thin Brick Walls," Western States Clay Products Association, San Francisco, CA, 1976.

10.4. ACI-SEASC Task Committee on Slender Walls, "Test Report on Slender Walls," ACI Southern California Chapter/Structural Engineers Association of Southern California, Los Angeles, CA, 1982, 125 pp.

10.5. Mayes, R.L., Clough, R.W., Hidalgo, P.A., and McNiven, H.D., "Seismic Research on Multistory Masonry Buildings, University of California 1972-1977," *Proceedings*, North American Masonry Conference, The Masonry Society, Boulder, CO, Aug. 1978.

CHAPTER 11—GLASS UNIT MASONRY

11.1—Scope

Glass unit masonry is used as nonload-bearing elements in interior and exterior walls, and in window openings. Code provisions are empirical, based on previous codes, successful performance, and manufacturers' recommendations.

11.3—Panel size

The Code limitations on panel size are based on structural and performance considerations. Height limits are more restrictive than length limits based on historical requirements rather than actual field experience or engineering principles. Fire resistance rating tests of assemblies may also establish limitations on panel size. Contact glass block manufacturers for technical data on the fire resistance ratings of panels, or refer to the latest issue of UL Building Materials Directory[11.1] and the local building code.

11.3.1 *Exterior standard-unit panels*—The wind load resistance curve[11.2,11.3] (Fig.11.3-1) is representative of the ultimate load limits for a variety of panel conditions. The 144 ft² (13.4 m²) area limit is based on a safety factor of 2.7 when the design wind pressure is 20 psf[11.4] (958 Pa).

11.3.2 *Exterior thin-unit panels*—There is no historical data for developing a curve for thin units. The Committee recommends limiting the exterior use of thin units to areas where the design wind pressure does not exceed 20 psf (958 Pa).

References

11.1 "Building Materials Directory," Underwriters Laboratories, Inc., Northbrook, IL, Product Category: Glass Block (KCJU), Current Edition.

11.2. "PC Glass Block Products," Installation Brochure (GB-185), Pittsburgh Corning Corp., Pittsburgh, PA, 1992.

11.3. "WECK Glass Blocks," Glashaus Inc., Arlington Heights, IL, 1992.

11.4. Smolenski, Chester P.,"A Study of Mortared PCC Glass Block Panel Lateral Load Resistance (Historical Perspective and Design Implications)," Pittsburgh Corning Corporation, Pittsburgh, PA, 1992.

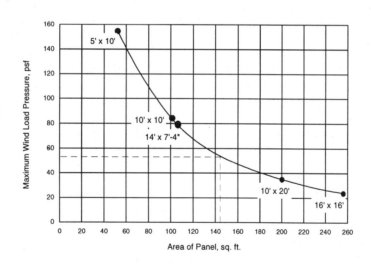

Example of How To Use Wind Load Resistance Curve: If using a design wind pressure of 20 psf (953 Pa), multiply by a safety factor of 2.7 and locate 54 psf (25586 Pa) wind pressure (on vertical axis), read across to curve and read corresponding 144 ft² (13.4 m²) maximum area per panel (on horizontal axis).

Fig 11.3-1—Glass masonry ultimate wind load resistance

CHAPTER 12—VENEERS

12.1—Scope

The traditional definition of veneer as an element without resistance to imposed load is adopted. The definition given is a variation of that found in model building codes. Modifications have been made to the definitions to clearly state how the veneer wythe is handled in design.

Implicit within these requirements is the knowledge that the veneer wythe transfers out-of-plane loads through the veneer anchors to the backing. The backing accepts and resists all anchor loads and is designed to resist all out-of-plane loads.

When utilizing anchored masonry veneer, the designer should consider the following conditions and assumptions:

a) The veneer may crack in flexure under service load.

b) Deflection of the backing should be limited to control crack width in the veneer and to provide veneer stability.

c) Connections of the anchor to the veneer and to the backing should be sufficient to transfer applied loads.

d) Differential movement should be considered in the design, detailing and construction.

e) Water will penetrate the veneer wythe and the wall system should be designed, detailed and constructed to prevent water penetration into the building.

f) Requirements for corrosion protection and fire resistance must be included.

The design of the backing should be in compliance with the appropriate standard for that material. Suggested standards are:

concrete - ACI 318, Building Code Requirements for Reinforced Concrete[12.1] American Concrete Institute

masonry - Chapters 1 through 10 of this Code

steel - Design for Cold-formed Steel Structural Members[12.2] American Iron and Steel Institute

wood - National Design Specification for Wood Construction[12.3] American Forest and Paper Association

If the backing is masonry and the exterior masonry wythe is not considered to add to the out-of-plane load resisting performance of the wall, the exterior wythe is masonry veneer. However, if the exterior wythe is considered to add to the load-resisting performance of the wall, the wall is properly termed a multi-wythe, non-composite wall rather than a veneer wall. Such walls are designed under Chapters 5 through 9 of this Code.

Manufacturers of steel studs and sheathing materials have published literature on the design of steel stud backing for anchored masonry veneer. Some recommendations have included composite action between the stud and the sheathing and load carrying participation by the veneer. The Metal Lath/Steel Framing Association has promoted a deflection limit of stud span length divided by 360[12.4]. The Brick Institute of America has held that an appropriate deflection limit should be in the range of stud span length divided by 600 to 720. The deflection is computed assuming that all of the load is resisted by the studs[12.5]. Neither set of assumptions will necessarily ensure that the veneer wythe remains uncracked at service load. In fact, the probability of cracking may be high[12.6]. However, post-cracking performance is satisfactory if the wall is properly designed, constructed and maintained with appropriate materials[12.7]. Plane frame computer programs are available for the rational structural design of anchored masonry veneer[12.6].

A deflection limit of stud span length divided by 200 times the specified veneer wythe thickness provides a maximum uniform crack width for various heights and various veneer thicknesses. Deflection limits do not reflect the actual distribution of load. They are simply a means of obtaining a minimum backing stiffness. The National Concrete Masonry Association provides a design methodology by which the stiffness properties of the masonry veneer and its backing are proportioned to achieve compatibility[12.8].

Masonry veneer with wood frame backing has been used successfully on one- and two-family residential construction for many years. Most of these applications are installed without a deflection analysis.

12.1.2 The Specifications were written for construction of masonry subjected to design stresses in accordance with the other chapters of this Code. Masonry veneer as defined by this Code is not subject to those design provisions. The Specifications Sections that are excluded cover materials and requirements that are not applicable to veneer construction or are items covered by specific requirements in this Chapter and are put here to be inclusive.

12.1.3 Adhered and anchored veneer definitions found in Section 2.2 are straightforward adaptations of existing definitions. See Figures 12.1-1 and 12.1-2 for anchored and adhered veneer, respectively.

Only anchored veneer is covered presently. Adhered veneer and stone veneer are not included in this Code at present except as a Special System of Construction, under Code Section 1.3.

12.2—General design requirements

Water penetration through the exterior veneer wythe is expected. The wall system must be designed and constructed to prevent water from entering the building.

The requirements given here and the minimum air space dimensions of Sections 12.8.3, 12.9.4 and 12.10.2 are those required for a drainage wall system. Proper drainage requires weepholes and a clear air space. It may be difficult to keep a 1 in. (25 mm) air space free from mortar bridging. Other options are to provide a wider air space, a vented air space or to use the rain screen principle.

Since there is no consideration of stress in the veneer there is no need to specify the compressive strength of the masonry.

12.3—Alternative design of anchored masonry veneer

There are no rational design provisions for anchored veneer in any code or standard. The intent of Section 12.3 is to permit the designer to use alternative means of supporting and anchoring masonry veneer. See Commentary Section 12.1 for conditions and assumptions to consider. The designer may choose to not consider stresses in the veneer wythe or may limit them to a selected value such as the

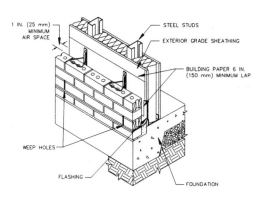

Fig. 12.1-1 Anchored veneer

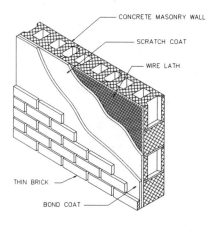

Fig. 12.1-2 Adhered veneer

allowable stresses of Chapter 6, the anticipated cracking stress, or some other limiting condition. The rational analysis used to distribute the loads must be consistent with the assumptions made. See Commentary Section 12.7 for information on anchors.

The designer should provide support of the veneer, control deflection of the backing, and consider anchor loads, stiffness, strength and corrosion; water penetration; and air and vapor transmission.

12.4—Prescriptive requirements for anchored masonry veneer

The provisions are based on the successful performance of anchored masonry veneer. These have been collected from a variety of sources and reflect current industry practices. Changes result from logical conclusions based on engineering consideration of the backing, anchor, and masonry wythe performance.

12.5—Vertical support of anchored masonry veneer

These requirements are based on current industry practice and current model building codes. Support does not need to occur at the floor level; it can occur at a window head or other convenient location.

The full provisions for preservative-treated wood foundations are found in the National Forest Products Association Technical Report 7[12.9].

There are no restrictions on the height limit of veneer backed by masonry or concrete, nor are there any requirements that the veneer weight be carried by intermediate supports. The designer should consider the effects of differential movement on the anchors and connection of the veneer to other building components.

12.7—Anchor requirements

It could be argued that the device between the veneer wythe and its backing is not an anchor as defined in the Code. That device is often referred to as a tie. However, the term anchor is used because of the widespread use of anchored veneer in model building codes and industry publications and the desire to differentiate from tie as used in other chapters.

U.S. industry practice has been combined with the requirements of the Canadian Standards Association[12.10] to produce the requirements given. Each anchor type has physical requirements that must be met. Minimum embedment requirements have been set for each of the anchor types to ensure load resistance against push-through or pull-out of the mortar joint. Maximum air space dimensions are set in Sections 12.8 through 12.10.

There are no performance requirements for veneer anchors in previous codes. Indeed, there are none in the industry. Tests on anchors have been reported[12.4, 12.11]. Many anchor manufacturers have strength and stiffness data for their proprietary anchors.

Veneer anchors typically allow for movement in the plane of the wall but resist movement perpendicular to the veneer wythe. The mechanical play in adjustable anchors and the stiffness of the anchor will influence load transfer between the veneer and the backing. Stiff anchors with minimal mechanical play provide more uniform transfer of load, increase the stress in the veneer and reduce veneer deflection.

The anchors listed in 12.7.6.1 are thought to have lower strength or stiffness than the more rigid plate-type anchors. Thus fewer plate-type anchors are required. These provisions may result in an increase in the number of anchors required when compared to the editions of the BOCA and SBCCI model building codes published in 1993[12.12, 12.13]. The number of anchors decreases in low seismic zones from the requirements in the UBC[12.14]. Anchor spacing is independent of backing type.

Anchor frequency should be calculated independently for the wall surface in each plane. That is, horizontal spacing of veneer anchors should not be continued from one plane of the veneer to another.

12.8—Masonry veneer anchored to wood backing

These requirements are similar to those used by industry and found in model building codes for years. The limitation on fastening corrugated anchors at a maximum distance from the bend is new. It is added to achieve better performance. The maximum distances between the veneer and the sheathing or wood stud is provided in order to obtain minimum compression capacity of anchors.

12.9—Masonry veneer anchored to steel backing

Most of these requirements are new, but they generally follow recommendations in current use[12.5, 12.15]. The minimum base metal thickness is given to provide sufficient pull-out resistance of screws.

12.10—Masonry veneer anchored to masonry or concrete backing

These requirements are similar to those used by industry and have been found in model building codes for many years.

12.11—Veneer laid in other than running bond

Masonry laid in other than running bond has similar requirements in Section 5.16. The area of steel required in 12.11 is equivalent to that in Section 5.16 for a nominal 4 in. (100 mm) wythe.

12.12—Requirements in seismic areas

These requirements provide several cumulative effects to improve veneer performance under seismic load. Many of them are based on similar requirements found in Chapter 30 of the Uniform Building Code[12.14]. The isolation from the structure reduces accidental loading and permits larger building deflections to occur without veneer damage. Support at each floor articulates the veneer and reduces the size of potentially damaged areas. An increased number of anchors increases veneer stability and reduces the possibility of falling debris. Joint reinforcement provides ductility and post-cracking strength. Added expansion joints further articulate the veneer, permit greater building deflection without veneer damage and limit stress development in the veneer.

References

12.1. Building Code Requirements for Reinforced Concrete, ACI 318-89 (revised 1992), American Concrete Institute, Detroit, MI, 1992.

12.2. Specification for the Design of Cold-Formed Steel Structural Members, American Iron and Steel Institute, August 10, 1986 Edition with December 11, 1989 Addendum, American Iron and Steel Institute, Washington, D.C., 1989.

12.3. ANSI/NFoPA National Design Specification for Wood Construction, American Forest & Paper Association, Washington, D.C., 1991.

12.4. Brown, R.H. and Arumula, J.O., "Brick Veneer with Metal Stud Backup - An Experimental and Analytical Study", Proceedings Second North American Masonry Conference, The Masonry Society, Boulder, CO, August 1982, pp. 13-1 to 13-20.

12.5. "Brick Veneer Steel Stud Panel Walls", Technical Notes on Brick Construction Number 28B Revised II, Brick Institute of America, Reston, VA, February 1987.

12.6. Grimm, C.T. and Klingner, R.E., "Crack Probability in Brick Veneer over Steel Studs", Proceedings Fifth North American Masonry Conference, The Masonry Society, Boulder, CO, June 1990, pp. 1323-1334.

12.7. Kelly, T., Goodson, M., Mayes, R., and Asher, J., "Analysis of the Behavior of Anchored Brick Veneer on Metal Stud Systems Subjected to Wind and Earthquake Forces", Proceedings Fifth North American Masonry Conference, The Masonry Society, Boulder, CO, June 1990, pp. 1359-1370.

12.8. "Structural Backup Systems for Concrete Masonry Veneers", NCMA TEK 16-3A, National Concrete Masonry Association, Herndon, VA, 1995.

12.9. "The Permanent Wood Foundation System", Technical Report No. 7, National Forest Products Association (now the American Forest and Paper Association), Washington, D.C., January 1987.

12.10. "Connectors for Masonry", CAN3-A370-M84, Canadian Standards Association, Rexdale, Ontario, Canada, 1984.

12.11. "Brick Veneer - New Frame Construction, Existing Frame Construction," Technical Notes on Brick and Tile Construction Number 28, Structural Clay Products Institute (now Brick Institute of America), Reston, VA, August 1966.

12.12. National Building Code, Building Officials and Code Administrators, Country Club Hills, IL, 1993.

12.13. Standard Building Code, Southern Building Code Congress International, Birmingham, AL, 1991.

12.14. Uniform Building Code, International Conference of Building Officials, Whittier, CA, 1991.

12.15. Drysdale, R.G. and Suter, G.T., "Exterior Wall Construction in High-Rise Buildings: Brick Veneer on Concrete, Masonry or Steel Stud Wall System", Canada Mortgage and Housing Corporation, Ottawa, Ontario, Canada, 1991.

Commentary on Specification for Masonry Structures (ACI 530.1-95/ASCE 6-95/TMS 602-95)

Reported by the Masonry Standards Joint Committee

James Colville
Chairman

Max L. Porter
Vice Chairman

J. Gregg Borchelt
Secretary

Maribeth S. Bradfield
Membership Secretary

Regular Members[1]:

Gene C. Abbate
Bechara E. Abboud
Bijan Ahmadi
Amde M. Amde
Richard H. Atkinson
William G. Bailey
Stuart R. Beavers
Robert J. Beiner
Frank Berg
Russell H. Brown
A. Dwayne Bryant
Kevin D. Callahan
Mario J. Catani
Robert W. Crooks
Kenneth G. Dagostino, Jr.

Gerald A. Dalrymple
Steve Dill
Russell T. Flynn
John A. Frauenhoffer
Thomas A. Gangel
Richard M. Gensert
Satyendra K. Ghosh
Clayford T. Grimm
John C. Grogan
Craig K. Haney
Gary C. Hart
Barbara Heller
Robert Hendershot
Mark B. Hogan
Thomas A. Holm

Rochelle C. Jaffe
John C. Kariotis
Richard E. Klingner
Walter Laska
L. Donald Leinweber
Hugh C. MacDonald, Jr.
Billy R. Manning
John H. Matthys
Robert McCluer
Donald G. McMican
George A. Miller
Reg Miller
Colin C. Munro
W. Thomas Munsell
Antonio Nanni

Joseph F. Neussendorfer
Joseph E. Saliba
Arturo Schultz
Matthew J. Scolforo
Daniel Shapiro
John M. Sheehan
Robert A. Speed
Ervell A. Staab
Jerry G. Stockbridge
Itzhak Tepper
Robert C. Thacker
Donald W. Vannoy
Terence A. Weigel
A. Rhett Whitlock

Associate Members[2]:

James E. Amrhein
David T. Biggs
James W. Cowie
John Chrysler
Terry M. Curtis
Walter L. Dickey
Jeffrey L. Elder
Brent A. Gabby

Hans R. Ganz
H. R. Hamilton, III
B. A. Haseltine
Edwin G. Hedstrom
A. W. Hendry
Thomas F. Herrell
Steve Lawrence
Nicholas T. Loomis

Robert F. Mast
John Melander
Raul Alamo Neihart
Robert L. Nelson
Rick Okawa
Adrian W. Page
Ruiz Lopez M. Rafael
Roscoe Reeves, Jr.

Phillip J. Samblanet
Richard C. Schumacher
John G. Tawresey
Robert D. Thomas
Dean J. Tills
Charles W. C. Yancey

SYNOPSIS

This Specification for Masonry Structures (ACI 530.1-95/ASCE 6-95/TMS-602-95) is written as a master specification and is required by the Code to control materials, labor and construction. This commentary discusses some of the considerations of the committee in developing this Specification with emphasis given to the explanation of new or revised provisions that may be unfamiliar to code users.

References to much of the research data used to prepare this Specification are cited for the user desiring to study individual items in greater detail. Other documents that provide suggestions for carrying out the provisions of this Specification are also cited. The Subjects covered are those found in this Specification. The chapter and article numbering of this Specification are followed throughout.

Keywords: clay brick; clay tile; concrete block; concrete brick; construction; construction materials; curing; grout; grouting; inspection; joints; **masonry**; materials handling; mortars (material and placement); quality assurance and quality control; reinforcing steel; **specifications**; tests; tolerances.

---

[1]Regular members fully participate in Committee activities, including responding to correspondence and voting.

[2]Associate members monitor Committee activities, but do not have voting privileges.

CONTENTS

INTRODUCTION

Chapter 3 of the "Building Code Requirements for Masonry Structures (ACI 530-95/ASCE 5-95/TMS 402-95)" makes the "Specification for Masonry Structures (ACI 530.1-95/ASCE 6-95/TMS 602-95)" an integral part of the Code. ACI 530.1/ASCE 6/TMS 602 Specification sets minimum construction requirements regarding the materials used in and the erection of masonry structures.

Specifications are written to set minimum acceptable levels of performance for the contractor. This commentary is directed to the Architect/Engineer writing the project specifications.

This commentary covers some of the points the Masonry Standards Joint Committee considered in developing the provisions of the Code which are written into this Specification. Further explanation and documentation of some of the provisions of this Specification are included. Comments on specific provisions are made under the corresponding chapter and article numbers of this Specification.

As stated in the Foreword, Specification ACI 530.1/ASCE 6/TMS 602 is a reference standard which the Architect/Engineer may cite in the Contract Documents for any project. Owners, through their representatives (Architect/Engineer) may write requirements into Contract Documents that are more stringent than those of ACI 530.1/ASCE 6/TMS 602. This can be accomplished with supplemental specifications to this Specification.

The contractor should not be asked through Contract Documents to comply with the Code or to assume responsibility regarding design (Code) requirements. The Code is not intended to be made a part of the Contract Documents.

The Foreword and Preface to Specification Checklist contain information that explains the function and use of this Specification. The Specification Checklist is a summary of the Articles that require a decision by the Architect/Engineer preparing the Contract Documents. Project specifications should include those items called out in the check list that are pertinent to the project. All projects will require response to the mandatory requirements.

PART 1—GENERAL

1.1—Summary

1.1C The scope of the work to be completed under this section of the Contract Documents is outlined. All of these tasks and materials will not appear in every project.

1.2—Definitions

For consistent application of this Specification, it is necessary to define terms which have particular meaning in this Specification. The definitions given are for use in application of this Specification only and do not always correspond to ordinary usage. The definition of the same term has been coordinated between the Code and Specification.

The permitted tolerance for units are found in the appropriate materials standards. Permitted tolerances for joints and masonry construction are found in this Specification. Nominal dimensions are usually used to identify the size of a masonry unit. The thickness or width is given first, followed by height and length. Nominal dimensions are normally given in whole numbers nearest to the specified dimensions. Specified dimensions are most often used for design calculations.

1.3—References

This list of standards includes material specifications, sampling, test methods, detailing requirements, design procedures and classifications. Standards produced by the American Society for Testing and Materials (ASTM) are referenced whenever possible. Material manufacturers and testing laboratories are familiar with ASTM standards which are the result of a consensus process. In the few cases not covered by existing standards the committee generated its own requirements. Specific dates are given since changes to the standard alters this Specification. Many of these standards require compliance with additional standards.

1.4—System description

1.4A *Compressive strength requirements*—Design is based on a certain f'_m and this compressive strength value must be achieved or exceeded. In a multi-wythe wall designed as a composite wall, the compressive strength of masonry for each wythe or grouted collar joint must equal or exceed f'_m.

1.4B *Compressive strength determination*

1.4B1 There are two separate means of determining the compressive strength of masonry. The unit strength method eliminates the expense of prism tests but is more conservative than the prism test method. The unit strength method was generated by using prism test data as shown in Fig. 1 and 2.

1.4B2 *Unit strength method*—Compliance with the requirement for f'_m based on the compressive strength of masonry units, grout and mortar type is permitted in lieu of prism testing.

The influence of mortar joint thickness is noted by the maximum joint thickness. Grout strength greater than or equal to f'_m fulfills the requirements of Specification Article 1.4A and Code Section 3.2.1.

1.4B2(a) *Clay masonry*—The values of net area compressive strength of clay masonry in Table 1 were derived using the following equation taken from Reference 1.1:

$$f'_m = A(400 + Bf_u)$$

where

A	=	1 (inspected masonry)
B	=	0.2 for Type N portland cement: lime mortar, 0.25 for Type S or M portland cement: lime mortar
f_u	=	average compressive strength of brick, psi (MPa)
f'_m	=	specified compressive strength of masonry

Rearranging terms and letting $A = 1.0$

$$f_u = \frac{f'_m - 400}{B}$$

These values were based on testing of solid clay masonry[1-1] units and portland cement-lime mortar. Further testing[1-2] has shown that the values are applicable for hollow clay masonry units and for both types of units with all mortar types. A plot of the data is shown in Fig. 1.

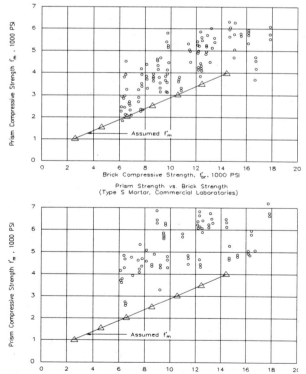

Fig. 1—Compressive strength of masonry versus clay masonry unit strength

1.4B2(b) *Concrete masonry*—In building codes[1.3, 1.4] prior to the Code, the compressive strength of concrete masonry was based on the net cross-sectional area of the masonry unit regardless of whether the prism was constructed using full or face shell mortar bedding. Furthermore, in these previous codes, the designer was required to base axial stress calculations on the net area of the unit regardless of the type of mortar bedding used. The Code has developed a standard compressive strength of masonry test procedure based on full mortar bedding of the prism. Strength calculations are based on dividing the maximum load on the prism by the net cross-sectional area of the masonry unit.

Design of concrete masonry sections is based on net cross sectional area which requires the designer to differentiate between face shell mortar bedded area and full mortar bedded area. The effect of these revisions changes the relationship between the unit compressive strength and the compressive strength of masonry to that listed in Table 2 in this Specification.

Table 2 lists compressive strength of masonry as related to concrete masonry unit strength and mortar type. These relationships are plotted in Fig. 2 along with data from 329 tests[1.5-1.11]. The curves in Fig. 2 are shown to be conservative when masonry strength is based on unit strength and mortar type. In order to use face shell bedded prism data in determining the Code unit strength to masonry compressive strength relationship used in the Code, a correlation factor between face shell prisms and full bedded prisms was developed. Based on 125 specimens, tested with full mortar bedding and face shell mortar bedding, the correlation factor was determined to be 1.29[1.5-1.7,1.12]. Face shell bedded prism strength multiplied by this correlation factor determines the full mortar bedded prism strength which is used in the Code.

1.4B3 *Prism test method*—The prism test method as specified by ASTM E 447 was selected as a uniform method of testing masonry to determine its compressive strength. The prism test method is used when the requirements of Article 1.4B2 do not apply or when otherwise required in the project specification.

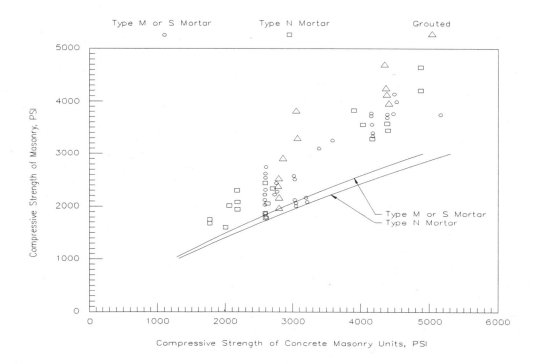

Fig. 2—Compressive strength of masonry versus concrete masonry unit strength

Compliance with the specified compressive strength of masonry can be determined by the prism method in place of the unit strength method. When compressive tests are required or desired, Method B of ASTM E 447 is to be used. This method uses the same materials and workmanship to construct the prisms as those to be used in the structure. However, several exceptions to Method B are listed based upon recent research and experience. References 1.13 through 1.17 discuss prism testing. Many more references on the prism test method parameters and results could be added. Several of these researchers have felt that more than one bed joint is needed in prisms, particularly for solid clay units. The simulation of full-scale masonry construction is thought to be better achieved by including several bed joints. Also, other factors affect the prism results, such as the prism-to-test-machine interface and the slenderness, therefore, a series of correction factors are provided, based upon h/t ratios as a means of providing a more uniform basis of comparison of the compressive strength.

The correction factor for clay masonry prism h/t ratio is from Reference 1.1 and is shown in Fig. 3.

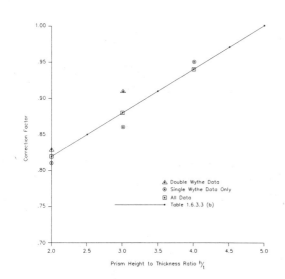

Fig. 3—Correction factors for clay masonry prisms

Article 1.4B3(c) requires that concrete masonry prisms shall have height-to-thickness ratios between 1.33 and 5.0. Fig. 4 compares the correction factors in Table 4 to test data.

These data show minor influence of slenderness on the strength of ungrouted prisms when compared to grouted prisms. However, the correction factors for slenderness of ungrouted prisms still apply[1.5, 1.10, 1.17-1.20].

1.5—Submittals

Submittals of materials, shop drawings and documentation, and their subsequent acceptance or rejection, on a timely basis will keep the project moving smoothly.

Submittals apply when the items mentioned are part of the project, or the project will be built during a time when the provisions are applicable.

Sample panels should contain the full range of unit and mortar color. All procedures, including cleaning and application of coatings and sealants, should be carried out on the sample panel. The effect of these materials and procedures on the masonry can then be determined before large areas are treated. Since it serves as a comparison of the finished work the sample panel should be maintained until all work has been accepted.

1.6—Quality assurance

The quality objective of the Owner will be met when the building is properly designed, completed using materials complying with product specifications and adequate construction practices and is adequately maintained. Both quality assurance and quality control requirements should be incorporated within the Contract Document. The extent of the quality assurance and the quality control program will vary with the size of the project. Success is dependent on open, but direct, communications between responsible parties within the design/build team. Records documenting the completion of the project are necessary to document the as-built condition.

(a) *Quality assurance*

Quality assurance consists of the actions taken by an Owner or the Owner's representative to provide assurance that what is being done and that what is being provided are in accordance with the applicable standards of good practice for the work. It is simply the administrative policies and responsibilities related to the quality control measures that will provide the Owner's quality objectives.

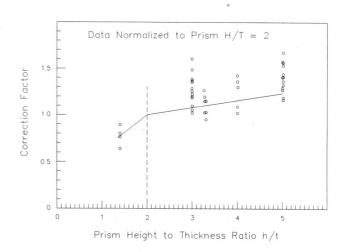

Fig. 4—Correction factors for ungrouted concrete masonry prisms

The quality assurance measures that shall apply to the Contract Documents, which include both the project specifications and the drawings, should address:

(1) organizational responsibilities,
(2) materials control,
(3) inspection,
(4) testing and evaluation,
(5) identification and resolution of noncomplying conditions,
(6) records.

The quality assurance considerations that may apply are delineated in Reference 1.20.

Organizational responsibilities delineate the responsibility and authority of the various personnel within the organization established. Direct line communications between on-site and off-site personnel should be well defined to allow interfacing of allied personnel.

Materials control should identify those required chemical and physical characteristics of individual materials required by the Contract Documents to be incorporated within the project. The controls should consider supplier evaluation and selection, procurement documents, inspection, delivery storage and handling, materials qualifications and records.

Inspection control should be established to assure that the masonry materials and construction practices comply with the requirements of the Contract Documents. Inspection personnel, the inspection program and inspection records should be addressed as inspection control measures. Testing and evaluation should be addressed in the quality assurance program. The program should allow for the selection and approval of a testing agency, which agency should be provided with prequalification test information and the rights for sampling and testing of specific masonry construction materials in accordance with referenced standards. The evaluation of test results by the testing agency should indicate compliance or noncompliance with a referenced standard.

Further quality assurance evaluation should allow an appraisal of the testing program and the handling of non-conformances. Acceptable values for all test methods should be given in the Contract Documents.

Identification and resolution of noncomplying conditions should be addressed in the Contract Documents. A responsible person should be identified to allow resolution of all nonconformances. In agreement with others in the design/ construct team, all resolutions should be either repaired, reworked, accepted as is, or rejected. Repaired and reworked conditions should initiate a reinspection.

Records control should be addressed in the Contract Documents. The distribution of documents during and after construction should be delineated. The review of documents should persist throughout the construction period to ascertain that all parties are informed and that records for documenting construction occurrences are available and correct after construction has been completed.

(b) *Quality control*

Quality control consists of the actions taken by a producer or contractor to provide control over what is being done and what is being provided so that the applicable standards of good practice for the work are followed. It is simply the systematic performance of construction, testing and inspection.

The specific requirements of the project specifications related to materials procurement and use should be implemented by quality control measures. Quality control measures may be dictated by the Owner's representative or be self-imposed by the responsible party charged with the conduct of a specific task associated with or contained in the Contract Documents.

Quality control measures essentially duplicate the quality assurance measures, except that the quality control measures apply to a specific project specification section or allied sections.

Members of the design/build team affected by quality control requirements include the designer, contractor and subcontractor, and inspection agency.

The responsible party performing the quality control measures should document the organizational representatives who will be a part of the quality control segment, their qualifications and the precise conduct during the performance of the quality control phase.

As an example of quality control measures, the general contractor may impose on the masonry subcontractor a request for a written procedure for cold weather masonry construction practices that will prevail.

As another example, the testing agency should be required to produce their laboratory accreditation document, essentially their quality assurance program. This document should indicate their organization, testing capabilities, qualification of personnel, test procedures, presence and calibration of physical test equipment, and records control.

1.6A1 The contractor establishes the source of supply of materials, mix designs and suggests change orders. Generally, requirements for selection of acceptable laboratory and for payment are in standard general conditions or other contractor documents provided by the Architect/Engineer. This article is to be used when no other documents exist.

1.6A2 The Code and this Specification require that all masonry be inspected.

The allowable stresses used in the Code are based on the premise that the work will be inspected, and that quality assurance and control measures will be implemented. As stated earlier in this section of the Commentary, the extent of these measures will vary with the complexity of the project. The Architect/Engineer, by employing various optional requirements in the body of the Standard Specification, can adjust the amount of quality assurance and control required. Good practice will determine the amount of inspection required. The method of payment for inspection services is usually handled in general conditions or other contract documents and usually will not be handled by this Article; hence the default statement.

The Owner and Architect/Engineer may require a testing laboratory to provide some or all of the tests mentioned. See also the Commentary for Articles 1.4 and 1.6.

Many of the requirements given in Article 1.6A4 may be carried out by other representatives of the Owner. The listing of duties of the Testing Agency and Contractor provide for a coordination of their tasks and a means of reporting results. All parties to the contract should maintain open communications.

The Contractor is bound by contract to supply and place the materials called for in the Contract Documents. Perfection is obviously the goal, but factors of safety included in the design method recognize that some deviation from perfection will exist. Engineering judgment must be used to evaluate reported deficiencies. Items which influence structural performance are controlled by the dimensional tolerances of Specification Article 3.3F.

1.7—Delivery, storage and handling

The performance of masonry materials can be lessened by contamination by dirt, water and other materials during delivery or at the jobsite.

Reinforcement and metal accessories are less prone to problems from handling than masonry materials.

1.8—Project conditions

1.8C *Cold weather construction*—The procedure described in this article represents the committee's consensus of current good construction practice and has been framed to generally agree with masonry industry recommendations[1.21].

The provisions of 1.8C are mandatory, even if the procedures submitted under Article 1.5B4a are not required. The contractor has several options to achieve the results required in Article 1.8C. The options are available because of the climatic extremes and their duration. When the air temperature at the jobsite or unit temperatures fall below 40° F (4.5° C) the cold weather protection plan submitted becomes mandatory. Work stoppage may be justified if a short cold spell is anticipated. Enclosures and heaters can be used as necessary.

Temperature of the masonry mortar may be measured using a metal tip immersion thermometer inserted into a sample of the mortar. The mortar sample may be mortar as contained in the mixer, in hoppers for transfer to the working face of the masonry or as available on mortar boards currently being used. The critical mortar temperatures are the temperature as sensed at the mixer and mortar board location.

Temperature of the masonry unit may be measured using a metallic surface contact thermometer.

It may be wiser for the contractor to enclose the entire area rather than make the sequential materials conditioning and protection modifications. Subsections (b) (1 through 4) apply to the work period when the masonry is actually being laid. Subsections (b) (5 through 8) apply more specifically to the period after the workday. Thus, ambient temperature conditions apply while work is in progress and mean daily temperatures apply to the time after the masonry is placed.

1.8D *Hot weather construction*—Temperature, solar radiation, wind, and relative humidity influence the absorption of masonry units, rate of set and drying rate. During periods of adverse drying of masonry construction it may be advisable to fog spray the wall at the end of the work day.

References

1.1. "Recommended Practice for Engineered Brick Masonry," Brick Institute of America (formerly Structural Clay Products Association), Reston, VA, 1969.

1.2. Brown, Russell H. and Borchelt, J. Gregg, "Compression Tests of Hollow Brick Units and Prisms," *Masonry Components to Assemblages, ASTM STP 1063*, J.H. Matthys, editor, American Society for Testing and Materials, Philadelphia, PA, 1990, p. 263-278.

1.3. ACI Committee 531, "Building Code Requirements for Concrete Masonry Structures (ACI 531-79) (Revised 1983)," American Concrete Institute, Detroit, MI, 1983, 20 pp.

1.4. "Specification for the Design and Construction of Load Bearing Concrete Masonry," (TR-75B), National Concrete Masonry Association, Herndon, VA, 1976.

1.5. Redmond, T.B., "Compressive Strength of Load Bearing Concrete Masonry Prisms," National Concrete Masonry Association Laboratory Tests, Herndon, VA, 1970, Unpublished.

1.6. Nacos, Charles J., "Comparison of Fully Bedded and Face-Shell Bedded Concrete Block," *Report* No. CE-495, Colorado State University, Fort Collins, CO, 1980, Appendix p. A-3.

1.7. Maurenbrecher, A.H.P., "Effect of Test Procedures on Compressive Strength of Masonry Prisms," *Proceedings*, 2nd Canadian Masonry Symposium, Carleton University, Ottawa, June 1980, pp. 119-132.

1.8. Self, M.W., "Structural Properties of Loading Bearing Concrete Masonry," *Masonry: Past and Present*, STP-589, ASTM, Philadelphia, PA, 1975, Table 8, p. 245.

1.9. Baussan, R., and Meyer, C., "Concrete Block Masonry Test Program," Columbia University, New York, NY, 1985.

1.10. Seaman, Jesse C., "Investigation of the Structural Properties of Reinforced Concrete Masonry," National Concrete Masonry Association, Herndon, VA, 1955.

1.11. Hamid, A.A.; Drysdale, R.G.; and Heidebrecht, A.C., "Effect of Grouting on the Strength Characteristics of Concrete Block Masonry," *Proceedings*, North American Masonry Conference, University of Colorado, Boulder, CO, Aug. 1978, pp. 11-1—11-17.

1.12. Hatzinikolas, M.; Longworth, J.; and Warwaruk, J., "The Effect of Joint Reinforcement on Vertical Load Carrying Capacity of Hollow Concrete Block Masonry," *Proceedings*, North American Masonry Conference, University of Colorado, Boulder, CO, Aug. 1978.

1.13. Atkinson, R.H., and Kingsley, G.R., "A Comparison of the Behavior of Clay and Concrete Masonry in Compression," Atkinson-Noland & Associates, Inc., Boulder, CO, Sept. 1985.

1.14. Priestley, M.J.N., and Elder, D.M., "Stress-Strain Curves for Unconfined and Confined Concrete Masonry," ACI JOURNAL, *Proceedings* V. 80, No. 3, Detroit, MI, May-June 1983, pp. 192-201.

1.15. Miller, D.E.; Noland, J.L.; and Feng, C.C., "Factors Influencing the Compressive Strength of Hollow Clay Unit Prisms," *Proceedings*, 5th International Brick Masonry Conference, Washington D.C., 1979.

1.16. Noland, J.L., "Proposed Test Method for Determining Compressive Strength of Clay-Unit Prisms," Atkinson-Noland & Associates, Inc., Boulder, CO, June 1982.

1.17. Hegemier, G.A.; Krishnamoorthy, G.; Nunn, R.O.; and Moorthy, T.V., "Prism Tests for the Compressive Strength of Concrete Masonry," *Proceedings*, North American Masonry Conference, University of Colorado, Boulder, CO, Aug. 1978, pp. 18-1—18-17.

1.18. Data from NCMA Prism Research Program, National Concrete Masonry Association, Herndon, VA, 1984, Unpublished Report.

1.19. Wong, Hong E., and Drysdale, Robert G., "Compression Characteristics of Concrete Block Masonry Prisms," *Masonry: Research, Application, and Problems*, STP-871, ASTM, Philadelphia, PA, 1985, pp. 167-177.

1.20. ACI Committee 121, "Quality Assurance Systems for Concrete Construction," (ACI 121R-85), American Concrete Institute, Detroit, MI 1985, 7 pp.

1.21. "Recommended Practices and Guide Specifications for Cold Weather Masonry Construction," International Masonry Industry All-Weather Council, Washington, D.C., 1973.

PART 2—PRODUCTS

2.1—Mortar materials

ASTM C 270 contains standards for all materials used to make mortar. Thus, component material specifications need not be listed. The Architect/ Engineer may wish to include only certain types of materials, or exclude others, to gain better control.

There are two methods of specifying mortar under ASTM C 270: proportions and properties. The proportions specification tells the contractor to mix the materials in the volummetric proportions given in ASTM C 270. These are repeated in Table C-1. The properties specification instructs the contractor to develop a mortar mix which will yield the specified properties under laboratory testing conditions. Table C-2 contains the required results outlined in ASTM C 270. The results are submitted to the owner's representative and the proportions of ingredients as determined in the lab are maintained in the field. Water added in the field is determined by the mason for both methods of specifying mortar. A mortar mixed by proportions may have the properties of a different mortar type. Higher lime content increases workability and water retentivity. ASTM C 270 has an Appendix on mortar selection.

Either proportions or properties, but not both should be specified. A good rule of thumb is to specify the weakest mortar that will perform adequately, not the strongest. Excessive amounts of pigments used to achieve mortar color may reduce both the compressive and bond strength of the masonry. Conformance to the maximum percentages indicated will limit the loss of strength to acceptable amounts. Due to the fine particle size, the water demand of the mortar increases when coloring pigments are used. Admixtures containing excessive amounts of chloride ions are detrimental to steel items placed in mortar or grout.

2.1B In exterior applications, certain exposure conditions or panel sizes may warrant the use of mortar type with high bond strength. Type S mortar has a higher bond strength than Type N mortar. Portland cement-lime mortars have a higher bond strength than some masonry cement mortars of the same type. The specified mortar type should take into account the performance of locally available materials and the size and

Table C-1—ASTM C 270 mortar proportion specification requirements

		Proportions by volume (cementitious materials)			
Mortar	Type	Portland cement or blended cement	Masonry cement M S N	Hydrated lime or lime putty	Aggregate ratio (measured in damp, loose conditions)
Cement-Lime	M	1	- - -	$\frac{1}{4}$	Not less than $2\frac{1}{4}$ and not more than 3 times the sum of the separate volumes of cementitious materials
	S	1	- - -	over $\frac{1}{4}$ to $\frac{1}{2}$	
	N	1	- - -	over $\frac{1}{2}$ to $1\frac{1}{4}$	
	O	1	- - -	over $1\frac{1}{4}$ to $2\frac{1}{2}$	
Masonry cement	M	1	- - 1	- - - - - -	
	M	-	1 - -	- - - - - -	
	S	$\frac{1}{2}$	- - 1	- - - - - -	
	S	-	- 1 -	- - - - - -	
	N	-	- - 1	- - - - - -	
	O	-	- - 1	- - - - - -	

Two air-entraining materials shall not be combined in mortar.

Table C-2—ASTM C 270 property specification requirements for laboratory-prepared mortar

Mortar	Type	Average compressive strength at 28 days, psi [1] (MPa)	Water retention min, percent	Air content max, percent	Aggregate ratio (measured in damp, loose conditions)
Cement-Lime	M	2500 (17.2)	75	12	Not less than $2\frac{1}{4}$ and not more than 3 times the sum of the separate volumes of cementitious materials
	S	1800 (12.4)	75	12	
	N	750 (5.2)	75	14 [2]	
	O	350 (2.4)	75	14 [2]	
Masonry cement	M	2500 (17.2)	75	-‡ [3]	
	S	1800 (12.4)	75	-‡ [3]	
	N	750 (5.2)	75	-- [3]	
	O	350 (2.4)	75	- [3]	

[1] Laboratory-prepared mortar only.

[2] When structural reinforcement is incorporated in cement-lime mortar, the maximum air content shall be 12 percent.

[3] When structural reinforcement is incorporated in masonry cement mortar, the maximum air content shall be 18 percent.

exposure conditions of the panel. Manufacturers recommend a waterproofing admixture or waterproof portland cement in exterior applications[2.1-2.3].

2.2—Grout materials

2.2A ASTM C 476 contains standards for all materials used to make grout. Thus, component material specifications need not be listed.

Admixtures for grout include those to increase flow and to reduce shrinkage.

2.3—Masonry materials

2.3A Concrete masonry units are made from lightweight and normal weight aggregate, water and cement. The units are available in a variety of shapes, sizes, colors and strengths. Since the properties of the concrete vary with the aggregate type and mix proportions, there is a range of physical properties and weights available in concrete masonry units.

Masonry units are selected for the use and appearance desired. Concrete masonry units are specified by grade, type and weight. Grade N units are for general use such as in exterior walls above or below grade that may or may not be exposed to moisture penetration or the weather or for exterior walls and backup. Grade S units are limited to use above grade on exterior walls with weather protective coatings and in walls not exposed to weather. Type I concrete masonry units are moisture controlled. Type II units are nonmoisture controlled units. There are three weight categories: normal, medium and lightweight, based on the density of the concrete used. Table C-3 summarizes the requirements found in the standards referenced.

ASTM C 744 covers the properties of units which have a resin facing on them. The units must meet the requirements of one of the other referenced standards.

2.3B Clay or shale masonry units are formed from those materials and referred to as brick or tile. Clay masonry units may be molded, pressed or extruded into the desired shape. Physical properties depend upon the raw materials, the method of forming and the firing temperature. Incipient fusion, a melting and joining of the clay particles, is necessary to develop the strength and

Table C-3—Concrete masonry unit requirements

ASTM specification	Unit name	Grade		
		Strength	Weight	Type
C 55	Concrete brick	yes	yes	yes
C 73	Sand lime brick	yes	no	no
C 90	Load-bearing	yes	yes	yes
C 129	Nonload bearing	yes	yes	yes
C 744	Prefaced	—	—	—

durability of clay masonry units. A wide variety of unit shapes, sizes, colors and strengths are available.

It is the intended use that determines which specification is applied. Generally, brick units are smaller than tile, tile is always cored, and brick may be solid or cored. Brick is normally exposed in use and most tile is covered. Grade, or class, is determined by exposure condition and has requirements for durability, usually given by compressive strength and absorption. Dimensional variations and allowable chips and cracks are controlled by type.

Table C-4 summarizes the requirements found in the standards referenced.

Table C-4—Clay brick and tile requirements

ASTM specifica-tion	Unit name	Mini-mum % solid	Grade		
			Strength	Absorption	Type
C 34	Load-bearing wall tile	a	yes	yes	no
C 56	Nonload-bearing wall tile	b	no	yes	no
C 62	Building brick (solid)	75	yes	yes	no
C 126	Ceramic glazed units	c	yes	no	yes
C 212	Structural facing tile	b	yes	no	yes
C 216	Facing brick (solid)	75	yes	yes	yes
C 652	Hollow brick	a	yes	yes	yes

Notes:

a—A minimum percent is given in this Specification. The percent solid is a function of the requirements for size and/or number of cells as well as the minimum shell and web thicknesses.

b—No minimum percent solid is given in this Specification. The percent solid is a function of the requirements for the number of cells and weights per square foot.

c—Solid masonry units-minimum percent solid is 75 percent. Hollow masonry units—no minimum percent solid is given in this Specification. Their percent solid is a function of the requirements for number of cells and the minimum shell and web thicknesses.

2.3C   Stone masonry units are most often selected by color and appearance. The referenced standards classify building stones by the properties shown in Table C-5. The values given in the standards serve as minimum requirements. Stone is often ordered by a particular quarry or color, rather than the classification method in the standard.

2.3D   Hollow glass masonry units are formed by fusing two molded halves of glass together to produce a partial vacuum in the resulting cavity. The resulting glass block units are available in a variety of shapes, sizes and patterns. Underwriters Laboratories inspects the manufacture and quality control operations of glass block production on a regular basis for UL approved units. The minimum face thickness is part of that inspection[2.4].

The block edges are usually treated in the factory with a

coating that can be clear or opaque. The primary purpose of the coating is to provide an expansion/contraction mechanism to reduce stress cracking and to improve the mortar bond.

Table C-5—Stone requirements

ASTM speci-fica-tion	Stone	Absorp-tion	Density	Com-pressive strength	Modu-lus of rupture	Abrasion resis-tance	Acid resis-tance
C 503	Marble	minimum	range	minimum	minimum	minimum	none
C 568	Limestone	range	range	range	range	range	none
C 615	Granite	minimum	minimum	minimum	minimum	minimum	none
C 616	Sandstone	range	range	range	range	range	none
C 629	Slate	range	none	none	minimum	minimum	range

2.4—Reinforcement and metal accessories

See Table C-6 for a summary of properties.

2.4D   *Stainless steel*—Corrosion resistance of stainless steel is greater than that of the other steels listed. Thus, it does not have to be coated for corrosion resistance.

2.4E   *Coatings for corrosion protection*—Amount of galvanizing required increases with severity of exposure[2.5-2.7]

2.5—Accessories

2.5A Movement joints are used to allow dimensional changes in masonry, minimize random wall cracks and other distress. Contraction (control) joints are used in concrete masonry to control the effect of shrinkage. These joints are free to open as shrinkage occurs. Expansion joints permit clay brick masonry to expand. Material used in expansion joints must be compressible.

Placement of movement joints is recommended by several publications[2.8-2.10]. Some general rules are: place at returns and jambs of wall openings, maximum spacing of 25 ft (7.6 m) or 3 times wall height. Typical movement joints are illustrated in Fig. 5. Shear keys keep the wall sections on either side of the movement joint from moving out of plane. Proper configuration must be available to fit properly. ASTM C 920 covers elastomeric joint sealants, either single or multicomponent.

Grade NS, Class 25, Use M is applicable to masonry construction. Expansion joint fillers must be compressible so the anticipated expansion of the masonry can occur without imposing stress.

2.5C   *Masonry cleaner*—Adverse reactions can occur with certain cleaning agents and masonry units. Hydrochloric acid has been observed to cause corrosion of metal ties. Care should be exercised in its use to minimize this potential problem. Manganese staining, efflorescence, "burning" of the units, white scum removal of the cement paste from the surface of the joints, damage to metals can occur through improper cleaning. The manufacturers of the masonry units should be consulted for recommendation of cleaning agents.

2.6—Mixing

2.6B   *Proportioning and mixing*—ASTM C 476 is strictly a proportion specification, with fine and coarse grout as the only choices. Their proportions are given in Table C-7.

The variation in the range of fine and coarse aggregates provides for adjustment required by aggregate type and gradation. As noted in Specification Table 5, the selection of the grout

Table C-6—Reinforcement and metal accessories

ASTM specification	Material	Use	Yield strength, ksi (MPa)	ASTM yield stress, MPa
A 36	Structural steel	Connectors	36 (248)	250
A 82	Steel wire	Joint reinforcement, ties	70 (483)	485
A 167	Stainless steel	Bolts, reinforcement, ties	30 (207)	205
A 185	Steel wire	Wire fabric, ties	75 (517)	485
A 307	Carbon steel	Connectors	60 (414)	
A 366	Carbon steel	Connectors	—	
A 496	Steel wire	Reinforcement	75 (517)	485
A 497	Steel wire fabric	Reinforcement, wire fabric	70 (483)	485
A 615	Billet steel	Reinforcement	40,60 (276, 414)	300,400
A 616	Rail steel	Reinforcement	50,60 (345, 414)	350,400
A 617	Axle steel	Reinforcement	40,60 (276, 414)	300,400
A 706	Low alloy steel	Reinforcement	60 (414)	

Table C-7—Grout proportions by volume

Grout type	Cement	Lime	Aggregate, damp, loose[1]	
			Fine	Coarse
Fine	1	0-$\frac{1}{10}$	$2\frac{1}{4}$ to 3	
Coarse	1	0-$\frac{1}{10}$	$2\frac{1}{4}$ to 3	1 to 2

[1]Times the sum of the volumes of the cementitious materials.

type depends on the size of the space to be grouted. Fine grout is selected for grout spaces with restricted openings. Coarse grout specified under ASTM C 476 has a maximum aggregate size which will pass through a ⅜ in. (9.5 mm) opening. Larger aggregate conforming to ASTM C 33, can be specified if the grout is placed in areas of unobstructed dimensions greater than 6 in.(152 mm).

Grout meeting the proportions of ASTM C 476 has typical compressive strengths shown in Table C-8 when measured by ASTM C 1019. Grout compressive strength is influenced by the water-cement ratio, aggregate content and the type of units used.

Table C-8—Grout strengths

Grout type	Location	Compressive strength, psi (MPa)			Reference
		Low	Mean	High	
Coarse	Lab	1965(13.5)	3106(21.4)	4000(27.6)	2.11
Coarse	Lab	3611(24.9)	4145(28.6)	4510(31.1)	2.12
Coarse	Field	5060(34.9)	5455(37.6)	5940(41.0)	2.13

Grout is proportioned by the volume of ingredients. Mixing requirements are not given in ASTM C 476, but job mixing time should be kept to a minimum. Grout is often ordered in ready-mix trucks rather than mixed at the jobsite. There are no mixing requirements which cover this procedure.

Since grout is placed in an absorptive form made of masonry units a high water content is required. A slump of at least 8 in. (203 mm) provides a mix fluid enough to be properly placed and supplies sufficient water to satisfy the water demand of the masonry units.

Small cavities or cells require grout with a higher slump than larger cavities or cells. As the surface area and unit shell thickness in contact with the grout decrease in relation to the volume of the grout, the slump of the grout should be reduced. Segregation of materials should not occur.

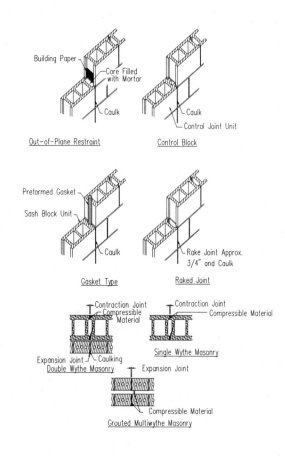

Fig. 5—Movement joints

The grout in place will have a lower water-cement ratio than when mixed. This concept of high slump and absorptive forms is different from that of concrete.

2.7—Fabrication

2.7A *Reinforcement*—These requirements have been industry standards for a long period of time.

2.7B *Prefabricated masonry*—ASTM C 901 covers the requirements for prefabricated masonry panels, including materials, structural design, dimensions and variations, workmanship, quality control, identification, shop drawings and handling.

References

2.1. "PC Glass Block Products," (GB-185), Pittsburgh Corning Corp., Pittsburgh, PA, 1992.

2.2. "WECK Glass Blocks," Glashaus Inc., Arlington Heights, IL, 1992.

2.3. Beall, Christine, "Tips on Designing, Detailing, and Specifying Glass Block Panels," Masonry Const., 3-89 ,pp 92-99.

2.4. "Follow-up Service Procedure", (File R2556), Underwriters Laboratories, Inc., Northbrook, IL, Ill.1, Sec. 1, Vol. 1, Current Edition.

2.5. Grimm, Clayford T., "Corrosion of Steel in Brick Masonry," *Masonry: Research, Application, and Problems*, STP-871, ASTM, Philadelphia, PA, 1985, pp. 67-87.

2.6. Catani, Mario J., "Protection of Embedded Steel in Masonry," *Construction Specifier*, V. 38, No. 1, Construction Specifications Institute, Alexandria, VA, Jan. 1985, p. 62.

2.7. "Corrosion Protection for Reinforcement and Connectors in Masonry," *NCMA TEK* 12-4, National Concrete Masonry Association, Herndon, VA, 1983, 4 pp.

2.8. Grimm, C. T., "Masonry Cracks: A Review of the Literature," *Masonry: Materials, Design, Construction, and Maintenance*, STP-992, ASTM, Philadelphia, PA, 1988.

2.9. "Movement," *Technical Notes on Brick and Tile* No. 18 and 18A, Brick Institute of America, Reston, VA, 1991.

2.10. "Control of Wall Movement with Concrete Masonry," *NCMA TEK* 10-2, National Concrete Masonry Association, Herndon, VA, 1972, 4 pp.

2.11. ACI-SEASC Task Committee on Slender Walls, "Test Report on Slender Walls," ACI Southern California Chapter/Structural Engineers Association of Southern California, Los Angeles, CA, 1982, 125 pp.

2.12. Li, D., and Neis, V.V., "The Performance of Reinforced Masonry Beams Subjected to Reversal Cyclic Loadings," *Proceedings*, 4th Canadian Masonry Symposium, Fredericton, New Brunswick, Canada, June 1986, V. 1, pp. 351-365.

2.13. Unpublished Field Test Report, File 80-617, B'Nai B'Rith Housing, Associated Testing Laboratories, Houston, TX, 1981.

PART 3—EXECUTION

3.1—Inspection

3.1A The tolerances in this Article are taken from Reference 3.1.

The dimensional tolerances of the supporting concrete are important since they control such aspects as mortar joint thickness and bearing area dimensions which influence the performance of the masonry. Tolerances for variation in grade or elevation are shown in Fig. 6. The specified width of the foundation is obviously more critical than its specified length. A foundation wider than specified will not normally cause structural problems.

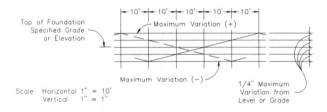

Fig 6—Tolerance for variation in grade or elevation

3.2—Preparation

3.2C *Wetting masonry units*—Concrete masonry units increase in volume when wetted and on subsequent drying will shrink. Clay masonry units with high absorption rates dry the mortar/unit interface. This results in lower bond which may result in a path of moisture intrusion.

3.2D *Debris*—Continuity in the grout is critical for uniform stress distribution. A clean space to receive the grout is necessary for this continuity. Inspection of the bottom of the space prior to grouting is critical to ensure that it is substantially clean and does not have accumulations of materials that would prevent continuity of the grout.

Placing the reinforcement prior to grouting prohibits holding the reinforcement by the stiffness of the grout and forcing the reinforcing bar into the grout. Loss of bond and misalignment of the reinforcement can occur if it is not placed prior to grouting.

Cleanouts can be achieved by removing the exposed face shell of units in hollow unit grouted masonry or individual units when grouting between wythes. The purpose of cleanouts is to allow the grout space to be adequately cleaned prior to grouting. They can also be used to verify reinforcement placing and tieing.

3.3—Masonry erection

Tolerances are established to limit eccentricity of applied load, and load carrying capacity of the masonry construction. Since masonry is usually used as an exposed material it is subjected to tighter dimensional tolerances than those for structural frames. The tolerances given are based on structural performance, not aesthetics.

The provisions for cavity width shown are for the space between wythes of non-composite masonry. The provisions do not apply to situations where masonry extends past floor slabs, spandrel beams, or other structural elements.

The remaining provisions set the standard for quality of workmanship and ensure that the structure is not over-loaded during construction.

3.4—Reinforcement installation

The requirements given ensure that:

a. galvanic action is inhibited,
b. location is as assumed in the design,
c. there is sufficient clearance for grout and mortar to surround reinforcement and accessories so stresses are properly transferred,
d. corrosion is delayed,
e. compatible lateral deflection of wythes is achieved.

Tolerances for placement of reinforcement in masonry first appeared in the 1985 Uniform Building Code[3.2]. Reinforcement location obviously influences structural performance of the member. Fig. 7 illustrates several devices used to secure reinforcement.

3.4D *Wall ties*—The Code does not permit the use of Z-ties in ungrouted, hollow unit masonry. The requirements for adjustable ties are shown in Fig. 8.

3.5—Grout placement

Grout may be placed by pumping, or pouring from large or small buckets. The amount of grout to be placed and contractor experience will influence the choice.

3.5B *Confinement*—Certain locations in the wall may not be grouted in order to reduce dead loads or allow placement of other materials such as insulation or wiring. Cross webs adjacent to cells to be grouted can be bedded with mortar to confine the grout. Metal lath, plastic screening, or other items can be used to plug cells below bond beams.

3.5C *Grout pour height*—Table 5 in the Specifications has been developed as a guide for grouting procedures. The designer can impose more stringent requirements if so desired. The recommended maximum height of grout pour corresponds with the least clear dimension of the grout space. The minimum width of grout space is used when the grout is placed between wythes. The minimum cell dimensions are used when grouting cells of hollow masonry units. As the height of the pour increases, the minimum grout space increases. The grout space dimensions are clear dimensions. See the Commentary for Section 3.1.2 of the Code for additional information.

Grout pour heights and minimum dimensions that meet the requirements of Table 5 do not automatically mean that the grout space will be filled. The top of a grout pour should not be located at the top of a unit, but within $1\frac{1}{2}$ in. (38 mm) of the bed joint.

3.5D *Grout lift height*—A lift is the height to which grout is placed into masonry in one continuous operation. After placement of a grout lift, water is absorbed by the masonry units. Following this water loss, a subsequent lift may be placed on top of the still plastic grout. If a lift of grout is permitted to set prior to placing the subsequent lift, then that height is the pour height.

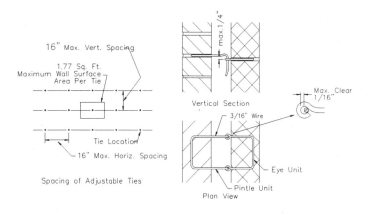

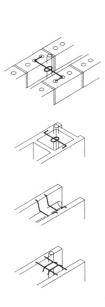

Fig. 7—Typical reinforcing bar position

This setting normally occurs if the grouting is stopped for more than one hour.

Grouted construction develops fluid pressure in the grout space. Grout pours composed of several lifts may develop this fluid pressure for the full pour height. The faces of hollow units with unbraced ends can break out. Wythes may separate. The wire ties between wythes may not be sufficient to prevent this occurrence.

3.5E *Consolidation*—Consolidation of grout is necessary to achieve complete filling of the grout space. Reconsolidation returns the grout to a plastic state and eliminates the voids resulting from the water loss from the grout by the masonry units. It is possible to have a height loss of 8 in. (203 mm) in 8 ft (2.4 m).

Consolidation and reconsolidation are normally achieved with a mechanical vibrator. A low velocity vibrator with a ¾ in. (19 mm) head is used. The vibrator is activated for one to two seconds in each grouted cell of hollow unit masonry. When double, open end units are used, one cell is considered to be formed by the two open ends placed together. When grouting between wythes, the vibrator is placed in the grout at points spaced 12 to 16 in. (305 to 406 mm) apart. Excess vibration does not improve consolidation and may blow out the face shells of hollow units or separate the wythes when grouting between wythes.

3.6A The frequency of testing given has long been an industry standard.

Fig 8—Adjustable ties

3.6B ASTM C 270 specifies mortar testing under laboratory conditions only for acceptance of mortar mixes under the properties specification. Field sampling and testing of mortar is conducted under ASTM C 780 and is used to verify consistency of materials and procedures, not mortar strength.

3.6C ASTM C 1019 requires a mold for the grout specimens made from the masonry units which will be in contact with the grout. Thus, the water absorption from the grout by the masonry units is simulated. Sampling and testing frequency may be based on the volume of grout to be placed rather than wall area.

References

3.1. ACI Committee 117, "Standard Specifications for Tolerances for Concrete Construction and Materials (ACI 117-90)," American Concrete Institute, Detroit, MI, 1981, 10 pp.

3.2 Uniform Building Code, International Conference of Building Officials, Whittier, CA, 1985.

INDEX